The Handbook of
Acoustic Bat Detection

The Handbook of Acoustic Bat Detection

Volker Runkel, Guido Gerding and Ulrich Marckmann

Translated by Iain Macmillan

PELAGIC PUBLISHING

Published by Pelagic Publishing
PO Box 874
Exeter
EX3 9BR
UK

www.pelagicpublishing.com

The Handbook of Acoustic Bat Detection

Originally published in German as
Handbuch: Praxis der akustischen Fledermauserfassung

ISBN 978-1-78427-220-3   *Paperback*
ISBN 978-1-78427-221-0   *ePub*
ISBN 978-1-78427-222-7   *ePDF*

A CIP record for this book is available from the British Library

Cover image: Kuhl's bat *Pipistrellus kuhlii* drinking. © Jens Rydell

Typeset by BBR Design, Sheffield

Printed and bound in India by Replika Press Pvt. Ltd.

Echolocation is sometimes just easier on the eyes

Contents

Preface and Acknowledgements

Since this book is a product of the authors' spare time, it has been longer in coming than originally planned. However, we hope that it has been well worth the wait! The aim of the second edition of this book, now under a new title, has been to incorporate suggestions from readers and to add content that did not make it into the first edition, published in German in 2018.

Acoustic bat recording has enjoyed an incredible and unprecedented increase in popularity among volunteers, scientists and professional writers of environmental reports. One of the reasons for this is the need for evidence when evaluating the effects on bats of new developments, above all wind farms. Only a few years ago, the use of data tended to be very unscientific and amateurish, but the huge number of very advanced devices now available on the market has opened up the prospect of all sorts of interesting and exciting projects. A new approach has become established, and has taken bat research and environmental impact assessments to a new level of sophistication.

Anybody involved with acoustic recording will be quick to admit that, despite all the euphoria, there are still many unanswered questions concerning the benefits and limitations of the new techniques. And there is a lack of clear guidelines on the processing of data. There are, for example, no clearly defined activity indices.

The aim of this book is to fill that gap and provide an overview of the applications of acoustic bat recording. It will not generally be possible to include exhaustive technical comparisons of recording equipment. The intention is rather to deal with some of the numerous questions relating to its practical use. The key technical concepts needed for working with ultrasound, and the scientific principles underlying it, are explained in the final chapter.

The authors have many years of experience with acoustic recording and the development of hardware and software for the recording and analysis of bat calls. We would like to thank the many people who, in numerous discussions, have directly or indirectly contributed to the development of this book. There are too many to name everybody, but we owe particular thanks to the following, in alphabetical order: A. Benk, L. Grosche, J. Koblitz and

U. Rahmel. I would also like to express my particular gratitude to Otto von Helversen, who made it possible for me, Volker Runkel, to devote myself to the world of bats.

I started working on bats and bioacoustics in 1996 at the University of Erlangen, in Professor Otto von Helversen's lab. We had access to tools that were not available elsewhere. Soon after I started my studies there, the first digital bat recorder with a high sampling rate and designed for field use was built by engineers at the university. Yet throughout most of those early years we were working with the standard bat detector, typical for Germany: a heterodyne detector. Nearly all bat workers in Germany used heterodyne systems. Thus, when talking about a bat detector we all had a clear picture of a device that has a frequency dial and makes the typical heterodyne noises.

We had read about the Anabat system, and knew at least one person working with it in Germany. We also had a simple frequency division detector. But it did not compare well with our heterodyne systems, so we never used it. Hence, the concept of frequency division was not part of our bat work.

With the availability of the university's real-time recording system, Anabat systems were also not interesting for our research. Time expansion detectors, mostly the Pettersson D240x, quickly dominated the bat worker market in Germany. From there it was only a small leap to the batcorder-system, which became the standard tool for consultants in Germany.

While finalising the translation of this book with the valuable help of Philip Briggs, I got new insights into how bat work is done in the UK – it is completely different to the situation in Germany. In the UK, frequency division is quite common and is used not only in actively searching for bats but also to record their calls for analysis. It is still used just as heterodyne detectors are in Germany; this is a method I have never used and was not much aware of. Kind of a cultural difference. Considering these fundamentally different techniques made it clearer to me how data is interpreted in the UK and which measurements are used for species identification and call description. One might think that all bat workers employ the same or similar approaches, but the history of detector techniques reveals how work is conducted in different places.

In this way, Philip has done great work to add some of these concepts to the book and thus raise the value of its content. Without him, certain chapters would have been misunderstood by many English-language readers. And all this with only small differences between our countries and with the same bat species – and indeed for some species even the same individuals!

Volker Runkel, March 2021

1 Acoustic recording of bats

The presence of bats in flight can easily be verified by acoustic methods, since they emit echolocation signals at regular intervals of between 2 and 20 times a second. Bats use these sounds and the resulting echoes to navigate and to find their prey (Griffin *et al.* 1960; Griffin 1995). These ultrasound signals are generally not audible to the human ear, and so a technical device, a bat detector, is needed to capture them. Nowadays, there is a wide variety of devices to make bat calls audible to the human ear or store them for later analysis. The technical differences between the devices determine their field of application and their reliability in the detection and identification of bat species. Not all devices are equally suitable for every task.

There is a large number of heterodyne, frequency division and time expansion detectors (for manual or hand-held operation) as well as direct sampling (also known as full spectrum) systems (for automated operation). A feature of the latter is that the sound is digitised directly and stored without modification. Descriptions of the technical details can be found in Chapter 14 (Section 14.3, *Heterodyne detectors*, onwards). Only some of the available solutions are ideally suited to autonomous operation (passive monitoring). Using a bat detection system is not always problem free, and the recording quality is sometimes not good enough for reliable identification of species, whether carried out manually or automatically.

This chapter discusses the potential applications of acoustic recording and briefly contrasts it with other surveying methods. This will enable the reader to gain an insight into the acoustic recording systems available, as well as a general understanding of the subject.

1.1 A survey of the technology

There are numerous technical aids for the acoustic recording of bat calls. These devices, known as bat detectors, are available in many different forms. The most popular types of device are introduced below to give an idea of the range available. A distinction must be drawn between *active* recording (human operators with a bat detector) and *passive* recording (automated monitoring).

1.1.1 Hand-held detectors for active recording

There are various hand-held detectors available to convert ultrasound to a pitch audible to the human ear. These will be heterodyne, frequency division or time expansion detectors (see Sections 14.3, *Heterodyne detectors*, 14.4, *Frequency division detectors*, and 14.5, *Time expansion detectors*) equipped with a loudspeaker or headphones. The different species of bat can be distinguished by their sound pattern and rhythm. Nowadays, there are also devices or attachments available for smart phones or tablets for detecting ultrasound which not only allow the acoustic reproduction of ultrasound, but also provide an optical image of the calls in the form of a real-time sonogram.

Heterodyne detectors are designed for instant identification in the field while the bat is present, while frequency division detectors and time expansion detectors are usually attached to a sound recording system so that recordings can be made for later call analysis using computer software (Meschede and Heller 2000; O'Donnell and Sedgeley 1994). In the early years, the time was recorded on tape at regular intervals using a clock with a sound signal. The whole technological apparatus was kept at a very simple level. Modern digital recorders store sounds with a time stamp in WAVE or MP3 format files.

In Germany and other countries heterodyne detectors are still used for passive monitoring by connecting them to a sound storage. Heterodyne detectors restrict tuning to a single frequency, and so can monitor only a limited call range around this frequency (±10 kHz). Species with calls outside this frequency range will be missed. This problem can, however, be overcome by the parallel use of two devices. Indeed, there have been two-channel devices available for some years which can monitor two separate frequency bands simultaneously. A user of a heterodyne detector can potentially identify a range of species based on clues such as the peak frequency, rhythm, repetition rate and tonal quality of the calls. Heterodyne detectors do not always allow precise identification of species, although it is usually possible to categorise calls by species group. With experience, some species, such as serotine, noctule and common, soprano and Nathusius' pipistrelles, can be fairly confidently identified. It should be emphasised that the application of these heterodyne devices for computer call analysis is limited by the fact that there can be no precise recording of the call frequencies (see also Section 5.1.1, *Identification by ear*); these detectors are principally designed for instant identification of the bat in the field while it is present through the use of audio clues (plus any visual clues available through observation of the bat).

For passive monitoring with heterodyne systems two parallel detectors are needed with frequency settings of around 25 and 40/45 kHz in order to obtain a representative result. Alternatively, two-channel devices with two independent frequency settings may be employed. This will ensure that all the important species groups are picked up, such as *Nyctalus* between 20 and 30 kHz, and *Pipistrellus* and *Myotis* at 40/45 kHz. However, there is the risk that the devices will not pick up the soprano pipistrelle, which emits calls at frequencies of 55 to 60 kHz.

Recordings from frequency division detectors are suitable for computer call analysis, though the resulting sonograms are less clear than from other broadband systems (time expansion and direct sampling detectors), which can make species identification more challenging and time consuming. Nevertheless, for a quick assessment of bat activity at a location on a single night, simple bat detectors of this sort are useful if no other technology is available.

Higher quality sonograms can be achieved through the use of a time expansion detector connected to a digital storage device (Section 14.5, *Time expansion detectors*). Like frequency division detectors, these time expansion detectors are broadband devices (recording across a wide frequency range which typically enables all frequencies in bat calls to be recorded and measured), but allow more effective identification of species, as the calls are recorded in a much greater level of detail. One disadvantage of the system is the down time caused by the tenfold slowing down of the internal recording during playback, which is necessary to allow storage by the WAVE recorder. A further recording can only be made when this is complete. Time expansion devices have, however, been rendered obsolete by direct sampling technology, and are now rarely used.

1.1.2 Detectors for passive automated recording

In contrast to manually operated devices, passive automated detectors do not necessarily have to make the bat calls audible to the human ear. They are designed primarily to store the sound data, and so have correspondingly different technical specifications. They operate autonomously and include a wide range of different technical designs which vary in their ability to distinguish between species. It is therefore necessary to include precise specifications of the technology and the settings used when documenting the results of a survey.

The latest development in automated acoustic recording of bats is the direct sampling (also known as full spectrum) detector. This stores calls digitally at a higher sampling rate (300 kHz or higher), recording ultrasound at its original frequencies rather than converting it to the human audible range, and allows for direct and automated further processing on the computer. With the appropriate power supply and weatherproofing, these devices can carry out continual acoustic monitoring over several months.

The Anabat system offers a similar solution. It does not store directly sampled sound data, but reduces the data by means of zero-crossing analysis after completion of the frequency division. Zero-crossing analysis works by counting the number of times the wave form crosses an imaginary line (the zero point) and from this calculates the frequency of the sound after every 8 or 16 crossings (depending on the data division setting selected on the detector). The data are then plotted on a frequency/time graph, rendering sound in the form of a series of data points. This tends to give a clearer representation of call structure than from a conventional frequency division system (Section 14.4) and the associated call analysis software offers a wide range of parameters for analysis and enables automated analysis through the application of filters. However, the zero-crossing process does not capture important information such as amplitude and harmonics.

1.2 Best practice

Acoustic methods have become a very popular and powerful tool in recent years, with the increasing availability of numerous recording systems and the automated analysis of calls (Brinkmann *et al.* 2011; Newson *et al.* 2014). More robust research tends to be characterised by the application of a mix of different methods (Hurst *et al.* 2015). As well as mobile and stationary acoustic detection, other established methods such as the use of nets, telemetry or monitoring of roosts should not be ignored. Often the parallel application of acoustic and non-acoustic methods will help with the collection of detailed information on bat activity and population structure. Only in this way can the influence of landscape changes on the local bat populations be meaningfully investigated.

All the available methods, acoustic and non-acoustic, have a bias towards particular species, and are thus mostly suitable only for certain limited investigations. Nevertheless, the most popular methods are understandably well established in bat research. Furthermore, new methods are constantly emerging, partly from combinations of existing solutions and partly from the development of completely new tools. All options need to be considered when planning a project, and a precise definition of the desired outcomes will allow the most suitable methods to be selected.

1.3 A survey of acoustic methodology

1.3.1 Automated recording – passive monitoring

This involves the use of systems which are set up in the field and record all bat calls autonomously (passively) for a given period of time. The sounds are recorded in such a way that they can be analysed on a computer. In general, heterodyne, frequency division and time expansion detectors are not used for autonomous recording, though there are exceptions (such as the Anabat SD2 with PDA).

Automated recording systems have to satisfy many requirements (Hayes 1997, 2000). In general, these systems should have the capacity to run unattended for several nights. Every bat flying in the vicinity should be clearly recorded, almost irrespective of where the bat is relative to the detector (omnidirectionality). The recordings need to be of a sufficient quality to allow automated and objective monitoring and identification of species.

Some systems are optimised for the automated and passive recording of bat activity over one or more nights. They can then operate autonomously and allow the simultaneous application of other methods by the bat workers on site. There are various automated systems available, each with its own advantages and disadvantages. Examples of such systems are the Elekon Batlogger, Wildlife Acoustics SM4BaT, ecoObs batcorder, the Avisoft system and Anabat. It is not the intention of this book to give a detailed account of the individual devices, but Chapter 9, *Criteria for detector systems*, provides a survey of the requirements and some comments on the available technology.

1.3.2 detectors for active recording

Simple detectors, consisting of a heterodyne or frequency division detector with a voice recorder or other recording device, are suitable for the sampling of bat activity for the duration of one night. They are convenient, but do not allow precise identification of species and heterodyne systems are limited in frequency bandwidth (see Section 1.1.1). In contrast, frequency division detectors are broadband detectors, making the recordings suitable for computer call analysis. This system, while relatively inexpensive, has the disadvantage of requiring a considerable amount of time for the analysis and identification of the calls.

1.3.3 Mobile acoustic recording

A hand-held detector (see Section 1.1.1) allows the recording of bat activity while on the move, and is particularly useful when doing transect surveys. It is then very easy to collect data from the whole of a survey area, whether on foot, by bike or by car. There are various protocols that can be followed. Activity may be recorded on the move or by stopping at regular intervals. The exact procedure must be adapted to the conditions on the ground and to the desired results. If it is not practicable to cover large areas on foot, it may make sense to use a bicycle, or if ultrasound noise is produced by walking through dense vegetation on a transect, it is sensible to make stops in order to avoid bat calls being obscured. The exact procedures must be established before the project begins. To allow meaningful comparisons between surveys of one site, the protocol must not be changed once it has been agreed. It must also be precisely documented so that, for example, different bat workers can record data in exactly the same way. In some circumstances, it may be appropriate to follow a recognised protocol on standards of recording.

The use of hand-held detectors is ideal for investigating the importance of landscape features for bats. When the bat worker is present in the habitat, invaluable empirical data can be collected by direct observation of the bats' behaviour. With a little experience, flight paths and potential roost areas can quickly be identified. Hand-held detectors are also helpful in recognising hotspots which can be monitored subsequently using passive techniques over periods of several days or weeks.

1.4 Non-acoustic methods – a digression

The following discussion of other commonly used methods of recording and researching bats is intended only as a brief overview. In order to achieve a scientifically comprehensive survey of bats, a mix of all available acoustic and non-acoustic methods should seriously be considered. Not all questions can be answered by acoustic methods alone.

1.4.1 Use of mist nets

For the assessment of the suitability of an area of habitat for bats, there are certain aspects which can be recorded only partially or not at all by acoustic methods. The capture of individual bats is essential, particularly for determining population structure and age and sex ratios, and also to identify certain species. Special mist nets are often used for this purpose. Mist netting does, of course, have its drawbacks, since some species can detect the net and avoid it. The method is sufficiently effective for many species and types of research in some locations, but it may not catch all species and will only represent a narrow time window. Since most species can be identified with 100% accuracy in the hand, however, the resulting data are of a high quality. With some experience and practice, the age and reproductive status can also be confidently determined.

In other circumstances, mist nets are less effective: for example, when the aim is to catch high-flying bats or bats whose echolocation system allows them to perceive the net and take evasive action. For these species, acoustic recording is a superior technique. All in all, a combination of modern automated detectors and mist netting is a good approach for many research projects (Murray *et al.* 1999; O'Farrell and Gannon 1999). The range of species caught in mist nets can be compared with the data from the automated detectors. Furthermore, it may allow a significant improvement in the acoustic identification of those species that are more difficult to recognise.

1.4.2 Telemetry

Telemetry is a method which can be used for the investigation of specific bats in their feeding areas (Wilkinson and Bradbury 1988) and locating roost sites, particularly of tree-dwelling species. Bats are captured and a small transmitter is attached. Regular monitoring by means of a receiver allows the recording of the successive locations of the radio-tagged animals. It is a very time consuming operation as the animal has first to be captured and then followed. For this reason, the technique is used very sparingly on just a few animals and for only a few nights in each case. It is an extremely useful investigative tool as it allows tracking of the movements of individual bats, and can be very helpful in locating roost sites. It may also be used for assessing the impact of landscape change if the bats are using the area for feeding. If experience is necessary for catching bats with mist nets, the same is even more true for telemetry. It is important that the attachment of a transmitter to a bat should not impede it or endanger its survival in any way. It should also be noted that not all bats can be tagged with a transmitter.

1.4.3 Monitoring of roost sites

An investigation into population changes can be carried out in some species by monitoring at their roost sites, using such techniques as emergence counts. Scientific investigations can use transponders and camera technology with daily measurements over a period of years to yield valuable datasets which allow the assessment of population changes. Anecdotal impressions and infrequent or irregular monitoring, as often happens with trees and bat boxes, will not provide data of sufficient quality.

Monitoring of roost sites is a very good technique for some species. The roosts of the greater mouse-eared bat in Germany, for example, are monitored regularly and emergence counts carried out within the framework of the EU Habitats Directive. In the case of species that change their roost site regularly and use a network of locations, such as the common pipistrelle, the method can be used, but it provides less reliable data on population change. This method is even more difficult with species that use hidden roost sites which are hard to find or observe.

In general, therefore, species that inhabit buildings and underground sites are much easier to investigate than those that live in woodlands and roost in trees or behind bark.

Winter roosts can also be monitored, although here the bats may be more hidden and thus more difficult to count. Many bats hang or hide in inaccessible places out of sight of the observer. Incomplete data of this sort may nevertheless allow an assessment of population changes if the observations are carried out systematically over a period of ten years or more over many roost sites, or are supplemented by the use of technology such as automated bat counters at roost entrances.

2 Examples of acoustic studies

There are numerous fields of application for the acoustic recording of bats. In this chapter some typical examples are introduced, with the emphasis on automated recording. Since long-term monitoring has come to prominence in the context of the planning of wind farms, this is dealt with in detail in Sections 2.4, *Long-term monitoring*, and 2.5, *Monitoring on the nacelles of wind turbines*.

2.1 Technical parameters

There are a large number of technical systems available, and in some cases they differ significantly from one another. When planning a study, it is important to define early on what is required from the recorded data, as this will determine which type of equipment is most appropriate (Waters and Walsh 1994; Adams *et al.* 2012). Variables such as the nature of the results sought, the project timeframe, the location of the study and many other factors will mean that not all technologies are equally suitable.

2.1.1 A comparison of techniques

The focus here will be primarily on automated recording, as this is the preferred option for many projects. An exhaustive comparison of the various detector systems is beyond the scope of this book, since such an exercise would involve the detailed examination of countless combinations and settings. However, some important aspects are compared with regard to the practical application of these systems and the validity of their results. The possible applications of acoustic recording are then explained in detail. In addition, Chapter 9, *Criteria for detector systems*, contains a further discussion of technical aspects. Other details on the pros and cons of manual and automated recording are presented fully in later chapters.

For a simple study of general bat activity at a single location on a single night, and without species identification, active recording with a hand-held device is often sufficient. However, a better recording system will be needed if accurate species data are required or if comparisons of activity are to be made between different sites or from the same site at different times.

The greater the need for precise identification, the higher the quality of the recording must be. Particularly if automated call analysis and identification are being used, which will be necessary when the number of recordings is large, a high-quality recording system should be employed. Only then will the recordings be of sufficient quality to allow automated call processing, and, if applicable, species identification.

In comparative studies, it is essential to ensure that the devices are as identical as possible in terms of sensitivity, in order to obtain comparable data. The recording sensitivity of the system needs to be known, and it must be programmable in all the devices used.

2.2 Data quality and its implications

In every study, the implementation of acoustic recording must be carefully considered. Locations, duration and settings of the devices used must be planned in line with the desired results. Before describing some typical recording procedures, it will be useful to look briefly at the 'standards' for various types of study (Table 2.1). The actual implementation will then usually be a blend of the simplified examples described here. There will, of course, always be situations that will deviate from the following recommendations. As with every technique, acoustic recording should be applied flexibly in order to achieve the necessary results.

Table 2.1   Summary of the application of automated acoustic recording in various types of study. The table gives guidelines, but in individual cases these should be adapted appropriately to the goals of the project.

| | Quantitative | Qualitative | Quantitative +
Qualitative |
| --- | --- | --- | --- |
| Sensitivity | High | Moderate | High (medium) |
| Duration | Several days/weeks | 2–3 nights | 2–3 nights/
long-term |
| Repetitions | Long-term/regularly | Results-oriented/
according to season | Regularly |

2.2.1 Quantitative data collection

The goal of quantitative data collection is to collect data which do not necessarily require the highest level of accuracy in terms of species identification. Rather they should allow as good an evaluation of the level of activity as possible. To achieve this, it is recommended that the recording device be set at high sensitivity, so maximising the recording range. A long recording period will furthermore compensate for any short-term fluctuations in activity.

Depending on the trigger algorithm and on the location, it may happen that non-bat calls, such as those of grasshoppers, are recorded, resulting in the data storage device being rapidly filled up over only a few nights.

Situations where quantitative recording is likely to be appropriate include long-term monitoring for the purposes of wind energy planning, or surveying bats around existing wind turbines. The calls are mostly identified only to family, with the primary aim being to classify the level of activity. In general, any recording which aims to ascertain the level of activity, without identifying precise species, can be defined as quantitative recording.

2.2.2 Qualitative data collection

The goal of qualitative data recording is to identify the range of species present at a location. It is vital to carry out recordings of as high a quality as possible, in order to pin down those species that are difficult to identify. If the recordings are of a poor quality, it makes the job of identification much harder. Since quantity is not relevant, the number of recordings is not so important. Therefore, it is a good idea to select only an intermediate level of sensitivity in the recording, except when dealing with bats with quiet calls. A high sensitivity setting should be chosen for these species, namely Bechstein's, greater mouse-eared, and all horseshoe and long-eared bats.

In order to record all the bat species in an area, suitable locations must be chosen carefully. If, for example, recording only takes place in open countryside, woodland species are likely to be partially or totally absent. Depending on the type and density of species anticipated at the location, it may be necessary to record over a period of a number of nights. By evaluating the results after each night, it may be possible to produce a saturation curve which will allow the recording schedule to be shortened. Examples of this can be found in Section 2.7, *Assessing biodiversity*.

2.2.3 Quantitative and qualitative data collection

If there is a need to record both quantitative and qualitative data, for example to study habitat use, it is advisable to set the device to medium sensitivity, except in some situations in open countryside where a high setting is better. Medium sensitivity will limit the recording to animals which are active in the study area, and exclude those which are outside, at a greater distance. Although there does not necessarily have to be long-term monitoring, it is nevertheless a good idea to record on several nights in succession, and to repeat this regularly. In this way it is possible to carry out successive studies of several locations without needing more than one or a few devices. However, recording over longer periods of time will always increase the validity of the study and ensure a reliable result (see also Section 10.1.1, *False negatives*, and Section 10.8, *The best activity index*).

2.3 Single-night surveys

In many studies, long-term monitoring in one location is not necessary. This is the case if fairly large areas are to be surveyed acoustically in a consistent way, primarily for the purpose of obtaining qualitative data. By regular monitoring, however, it is possible to obtain semi-quantitative results too.

Such surveys can be carried out either passively or actively. An experienced bat worker can investigate activity and the range of species quite effectively by walking the transect with a hand-held device. Alternatively, a passive recording device can be set up for a period of one or two nights in a location to investigate night-time activity and the range of species present. If various locations are being monitored using the same technology and the same settings, comparisons can be made between the sites. The optimal mode of operation really depends on the nature of the study. Often these types of time-limited study will be undertaken in conjunction with other methods. They are particularly suitable when constant use of the site by bats can be assumed. On the other hand, if the study is related to short-term phenomena such as bats migrating, exploring new roost sites, or exploiting sporadically occurring sources of food, then this must be monitored by regular or constant recording. The following section provides a rough breakdown of the best methods for each type of study.

2.3.1 Recording of general activity

A typical application of automated detectors is in the recording of activity at a single location for qualitative purposes. The operation will be repeated several times over the year, with the frequency varying according to the study. The range of species, and possibly also the frequency of use of the site by bats, can thus be determined (semi-quantitative data). In order to ensure higher-quality, echo-free recording, it is best to set up the devices on open ground, away from landscape features, to allow 360° monitoring. Furthermore, a much better set of data will be achieved by recording at a site over several nights rather than on just one, so that short-term variation and minor activity patterns can be eliminated. The sensitivity of the device should be on a medium setting.

If the primary goal is to study species such as noctules which fly high when in the open, a sensitivity setting in the upper range is preferred (despite the fact that noctules use loud calls, they fly not only very high but often also direct their beam forward and not downward, thus limiting detection range). A high sensitivity should also be used for species that may fly within closer range of the detector's microphone but have very quiet calls, such as brown long-eared bat and barbastelle. For devices which have a limited recording duration, the recording time should be set to between one and five seconds. In this way, activity indices can better be calculated for the comparison of data (see Section 10.2, *Quantification of activity – using identical recording systems*).

Recording on a few single nights over the year has the disadvantage that it will not be possible to provide an uninterrupted record of activity. Bats are highly mobile and to a certain degree exhibit very opportunistic behaviour. Some species change their hunting areas over the year as often as every few days, and thus will not occur with the same intensity all the time at one particular site. For this reason, it is not possible from several single-night surveys over a year to make a comprehensive assessment of the activity of a species. The absence of certain species should not be over-interpreted, since they have perhaps only been 'missed' (see also Section 10.1.1, *False negatives*, and Section 10.8, *The best activity index*). It follows then that the results should only be interpreted with reference to the bats recorded, unless the data were collected continually over a long period of time (see also Section 2.7, *Assessing biodiversity*).

2.3.2 Flight paths

One particular variant of general activity monitoring is the study of flight paths. Automated recording systems can be of use here, but, as a rule, flight paths are best found with a hand-held detector (heterodyne or frequency division), which gives immediate acoustic feedback as the bats pass. Flight paths can be determined both acoustically and optically as the bat worker moves along the transect of the study area. This is particularly effective immediately after emergence, as the bats fly along the landscape features at regular intervals in the same direction. It is really only by direct human observation of flight behaviour that a flight path can be accurately mapped.

Automated recording can nevertheless provide data for a comparison of activity between flight paths or between recording dates. If several recording devices are set up along an already identified flight path, a time comparison of the recordings can be made to supplement the manual observations. The recording device should be set at medium sensitivity if the site is directly on the flight path, but otherwise a high sensitivity setting will be more useful. The setting should also be adapted to the type of echolocation and flight style of the species concerned. Here the automated recording can be supplemented by simultaneous visual observation at the various locations.

2.3.3 Wildlife crossing structures

As with flight paths, structures to help bats in crossing roads, such as green bridges, can be monitored with automated recording devices (Abbott *et al.* 2012; Berthinussen and Altringham 2012). In order to evaluate the resulting data meaningfully, it is necessary to make a distinction between at least two different scenarios. Established structures are easier to monitor. Ideally a recording device is installed at each end of a crossing. Direct comparison of the recording times will then enable the surveyor to establish whether and in which direction a crossing has occurred. If only one device is available, this can be placed in the middle of the crossing structure. In the case of underpasses and wildlife tunnels, the recording quality is likely to be inferior because of interference from echoes. In such cases, a medium to low sensitivity setting on the device will be more appropriate.

Newly installed crossing structures must be monitored to see whether bats are in fact approaching them. An absence of animals crossing may be due not only to an unsuitable structure, but also to the fact that the bats are not even flying up to them. This may be the case when there is a lack of suitable guiding structures in the landscape such as hedgerows on the flight approach side. In such situations, the flight approach side should always be monitored in order to establish the presence of the target species.

This type of monitoring study should be carried out immediately after the landscape modifications and the installation of the wildlife crossing structures, and for as long as possible. Initially the frequency of usage will be low, as the bats become familiar with the changes, and it will be necessary to record on several nights. Once the data show that bats are using the structures more often, monitoring frequency can be reduced.

2.4 Long-term monitoring

The long-term monitoring of bats is most easily carried out using automated acoustic methods. Sometimes it is the only applicable method, as is the case with the monitoring of wind farms, which is dealt with in Section 2.5, *Monitoring on the nacelles of wind turbines*. Here it is best to install a detector which operates for the entire duration of the activity period. That would mean uninterrupted acoustic recording for six to nine months from March/April to October/November. Such studies are frequently carried out as part of the planning process for wind turbines. The detector is sometimes installed on a mast or in the top of a tree. Alternatively, a modified form of this technology may be mounted on the nacelle of the wind turbine. This method allows the recording of the range of species and their activity patterns over the course of the year, part of which may include migration. It provides quantitative and, if required, qualitative data. It is only with this level of monitoring that authoritative statements can be made about the threat to bats from the operation of wind turbines.

In order to allow such long-term operation, it will first be necessary to ensure that the power supply is equal to the task. Secondly, there must be regular daily or weekly checks to ensure that the microphone is operating correctly.

2.4.1 Location

Since the device is installed in one place for a period of several months and has to provide representative data for the study area, the location must be carefully selected. It needs both to provide adequate coverage of the study area, and to be suitable for the study aims. The device must not be placed too close to structures which have above- or below-average activity, as this could skew the data. Features which act as guiding structures on bat flight paths are better monitored by supplementary mobile recorders.

A location on open ground at a distance of between 10 and 15 m from other structures is often a good solution, as it will capture both the species that hunt in the open and those which prefer to skirt along buildings and vegetation. It will, however, fail to pick up bats that spend most of their time hunting in woodland. It is recommended that the device is installed on a mast at a height of between 4 and 8 m in order to protect it from theft.

During the preliminary planning for wind turbines in woodland, it is essential to record both below and above the treetops, as both high-flying and woodland bats will be affected by such developments. Regular or constant recordings must therefore be made, both in the middle of the stands of trees and on woodland paths. For the measurement of activity above the treetops, the construction of a sufficiently high mast is almost indispensable, because the detector should be installed at least 5–10 m above the crowns of the trees. Studies involving the use of balloons or kites generally only provide anecdotal evidence, and are not an adequate substitute for long-term monitoring. If there is no wind-measuring mast available, it is possible to use a location in a large clearing of between 30 and 60 m in diameter and still record the species that hunt in the open above woodlands. The recording device needs to be installed as high above the ground as possible. The device could perhaps be attached to a purpose-built structure in the clearing. Alternatively, a suitable tree on the edge of the clearing may be used, as long as the microphone is not screened off by any branches in the crown (Figure 2.1).

For long-term monitoring, the location must be appropriate for the study in question. Decisions must be made regarding quantity of data and the required dataset so as to obtain the desired outcome. Long-term acoustic monitoring is not always appropriate for some studies. For the bats with quieter calls, such as long-eared bat species in woodland, it is unusual to obtain good recordings. In these and other similar situations, long-term acoustic monitoring does not provide sufficient data for assessing the patterns of habitat use.

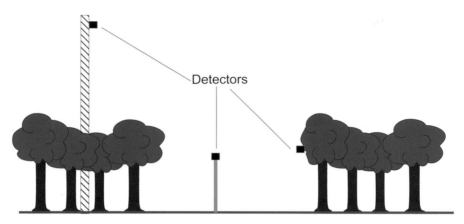

Figure 2.1   Options for long-term recording of species flying in open woodland areas.

It is not only the precise location of the recording device, but also the number of devices, that needs to be adapted to the study undertaken. Often a single recording device is sufficient, above all when in-depth accompanying studies are being undertaken. Several locations will, on the other hand, be indispensable when comparing habitats or studying very cluttered environments.

2.4.2 Settings for long-term monitoring

For long-term monitoring, high sensitivity is necessary to achieve the required range. However, only those sounds should be recorded that can be analysed. If the trigger sensitivity is set too high, it can under certain circumstances also lead to much time being wasted processing unusable low-volume calls. A judgement must also be made regarding appropriate frequency parameters of the trigger. In many locations, particularly in the summer, trigger settings with the frequency set too low will lead to large quantities of data being generated by bush-crickets or birds, and to a lot of time wasted in the subsequent analysis. Finally, recording should always include as a minimum the period from sunset to sunrise.

2.4.3 Long-term recording

The device must record data over the entire period of activity without any gaps. As Section 2.7, *Assessing biodiversity*, shows, recording sequences of more than 100 days are sometimes necessary to pick up the full range of species. In order not to miss short-term events, such as autumn migration, there is a need to cover all times of the year which are of significance to bats.

The electricity supply and data storage must be designed to allow long operating periods without servicing of the equipment. Our experience shows that, on average, up to 300 recordings per night must be expected, and so a 16 GB card will allow an operating time of at least a month and possibly up to three months.

During the recording, checks should be made regularly to ensure that the device, particularly the microphone, is operating correctly. Failure to do this could mean a loss of data and a need for the whole process to be repeated, with all the costs that this involves for the contractor. When setting up such an arrangement, it is important to have a system which carries out a daily test of the state of the device and the correct functioning of the microphone, and submits a daily report to the user by SMS or other messaging system.

2.4.4 Supplementary surveys

Long-term monitoring should be supported and enhanced by transect surveys on foot. By carrying out one to two weeks of all-night surveys, bat workers will be able to identify activity hotspots, flight paths and roost sites which are not identifiable using long-term passive monitoring. A further technique is to catch bats with mist nets, which is a useful method for studies requiring information on the sex ratios of the bats or on species of bats that are difficult to identify acoustically. The extent to which these supplementary techniques are used will depend largely on the nature of the study area and the aims of the project.

2.5 Monitoring on the nacelles of wind turbines

This is a particular variant of long-term recording, carried out on the nacelle of a wind turbine, and has been widely practised in many European countries since the early 2000s. Initially it was only used in a few individual projects, but it now belongs to the standard methods within the framework of the licensing procedures for wind turbines. It was used systematically from 2007 on 72 wind turbines as part of the RENEBAT project in Germany (Brinkmann *et al.* 2011). Searches for bats killed by wind turbines, which were carried out simultaneously with acoustic recording, allowed the correlation of the two sets of data. This project led to nacelle monitoring being adopted into the species protection guidelines of many of the German federal states. It has become the standard method for determining the shutdown periods for wind turbines as a mitigation measure to reduce bat deaths.

This type of monitoring entails the installation of an automated detector, generally a direct sampling device, in the nacelle. It is normal practice to install the microphone slanting downwards in the casing of the nacelle. The detector itself remains in the nacelle. It would be beneficial for any such project to use exactly the same methodology as the RENEBAT project, which is the largest study of bat casualties caused by wind turbines. Only by using the identical procedures will the data be comparable. Although there are no overall European guidelines, this is already happening in many

European countries, such as France, Austria, Belgium and the Netherlands. In Germany, the guidelines of the German federal states always refer to the project and require identical application of the technology. By setting up detectors using the RENEBAT methodology, it will be possible to use ProBat software to model casualties and calculate curtailment algorithms to halt the turbines when bats enter the danger area.

One of the main problems in many of the guideline frameworks has been that, for a long time, the precise settings for the devices have not been explicitly included in the wording. There has only been a reference to the RENEBAT project, and other information has been completely lacking in the documentation. Correct settings are absolutely critical for the comparability of the data.[1] Incorrect settings nearly always result in an underestimation of bat activity. This leads to legal uncertainties, because of the inadequate protection afforded to the bats. In the RENEBAT project, bat activity was measured by the *number of recordings*. The number of recordings is very much dependent both on how the recordings are made (whether it is done with pauses, fixed or variable recording lengths, etc.) and on the technology employed. In the RENEBAT project, very short recordings were produced because of the post-trigger settings of 200 ms (batcorder). Other post-trigger settings and different technologies may only result in longer recordings, which will probably lead to a significantly lower evaluation of activity (see also the example of the Leisler's bat in Section 10.2.1, *The number of recordings*).

In order to ensure comparable data, the only recording devices that should be used are those which operate in an identical way to the ones employed in the RENEBAT project, namely the Anabat SD2, Avisoft USG (and thus also the BATmode-System). The Batlogger, SM2Bat, SM3Bat, SM4Bat, Batomania Horchbox and the D500X were all considered as unsuitable for the project. None of these devices was calibrated for sensitivity, one of the main prerequisites for estimating the number of bats killed by collision with wind turbines. Because of its unreliability in recognising bat calls, the SM2Bat, for example, was rejected in unpublished tests as completely unsuitable. In the guidelines of the German Länder (states), there is no further discussion of the technical suitability, and reference is usually made just to direct sampling systems in general. In cases of doubt, however, a legally compliant recording is only possible under certain circumstances. A more detailed discussion of the aspects of triggering systems of the various detectors is to be found in Section 9.4, *Triggering systems*.

---

1   The only exception is the Bavarian Ministry for the Environment, which included this information in the FAQs of its guidelines.

If data have been collected in line with the RENEBAT methodology, data should be analysed using the software tool ProBat,[2] which was developed by the University of Erlangen-Nuremberg from the results of this project. It offers the only way of using a statistical model to convert the acoustic data into figures for collision mortality. In contrast to other techniques, this allows the calculation of precise statistics. The operation of the turbines can be modified so that the number of bats killed falls below stipulated threshold levels, in line with EU and Natura 2000 legislation. If ProBat software is used, there will be numerous requirements to be fulfilled over and above those of the particular hardware used.

A discussion of the advantages of the method and suggestions for possible improvements are featured in Chapter 12, *Nacelle monitoring – its benefits and limitations*.

2.6 Roost site monitoring

When searching for new roost sites, or protecting known ones, for example during building work, it can be helpful to make use of long-term acoustic recordings to provide valuable information about the nature and extent of the activity of the bats. In this way, potential roosts, including sites in and around buildings, can be monitored for actual use by bats. When planning this type of survey, it has to be borne in mind that the analysis of the calls will be more difficult, because bats near their roost often use calls that are shorter and more frequency modulated than their typical calls, and calls which are still insufficiently documented. Although automated analysis will correctly identify individual sequences, the proportion of incorrectly identified or unidentified calls will be very high. Manual analysis of recordings will be indispensable in these situations. In roosts with large numbers of bats, there may be huge quantities of data to analyse each night.

Acoustic recording methods lend themselves to the monitoring of activity at roost sites in buildings in which there is already some knowledge of the location and type of roost. The effectiveness in determining activity patterns increases as knowledge of the site improves. It must be stressed, however, that nothing can be said about the number of bats present. When investigating roosts where swarming occurs, acoustic recording is generally only suitable under certain circumstances. It is not possible to obtain important information on sex ratios and breeding condition, or on the number of individuals and the frequency with which each individual uses the roost. Acoustic recording only really allows a general indication of the level of activity.

2   http://windbat.techfak.fau.de/index.shtml

2.6.1 Location

As with the location of camera traps or light barriers, the ideal position for recording is near to where the bats fly in and out of the roost. However, in comparison with optical methods, the device should be placed further away to avoid calls from inside the roost being recorded. It is even difficult to distinguish between the calls of bats entering and leaving the roost. It is therefore better to place the microphone at a distance. In roosts with low numbers of bats this need not be very far because it can be assumed that there will be only a small amount of swarming activity and correspondingly few social calls. In roosts with large numbers, on the other hand, and particularly where there is lively swarming activity, a greater distance must be selected.

The recording can, of course, be carried out near or even within the roost, but then a large proportion of the recordings may have to be carried out with hand-held devices. In some studies, this may have additional advantages, such as when measuring winter activity in a roost, or determining the use or non-use of potential roost sites.

2.6.2 Settings

For roost surveys, it is recommended to use a setting of intermediate to low sensitivity, which will record bats close to the microphone. This will give priority to calls that can be automatically processed and will produce a manageable number of recordings. If the recording range is too great, quieter calls will be picked up, and with large roosts this can lead to an almost constant stream of recordings. With several thousand recordings, this will quickly result in large quantities of data (8–12 GB or more) in a single night.

The recording period should at the very least always include the time between sunset and sunrise. A longer operating period is advisable so as not to miss early emergences or late returns.

2.6.3 Supplementary investigations

Roost monitoring is only really of scientific value if the species of bat are identified by capture or internal site checks. This is indispensable if it is one of the long-eared species or the whiskered or Brandt's bats, which cannot be reliably identified by acoustic means. Similarly, it is the only way to determine if the site is a maternity or other type of roost.

2.7 Assessing biodiversity

When making an initial assessment of biodiversity, simple, quick and reproducible methods are used to determine the range of species in a habitat. Small errors are not significant when dealing with groups with large numbers of species, as it is more a matter of comparing the biodiversity of several locations. Acoustic recording is well suited to determining the variety of bat species. However, because of the opportunistic behaviour patterns of bats, it is important to be aware that recording on a small number of nights will not provide a comprehensive picture of all the species using the habitat. Some species are difficult to identify by acoustic methods, whether it be because of the quiet calls of Bechstein's, long-eared or horseshoe bats, or because of the unreliability of identifying calls when only a few sequences are picked up. Therefore, the results of acoustic surveys should be supplemented by other methods, particularly when dealing with rare or especially endangered species that are difficult to identify.

With diversity surveys, irrespective of the method used, it is important to determine how many sampling operations will be needed to determine the range of species present. Although there are various techniques for extrapolating the total number of species from a sampling survey, they are not reliable when applied to areas such as central Europe where there is a relatively small number of bat species present. As a rule, it is also important to identify not just the number of species, but also the precise species, as it is only then that it is possible to design and coordinate interventions or measures, or implement statutory requirements, such as species-by-species evaluation.

2.7.1 The scope of an investigation

How many recordings are needed at a location to investigate the range of species will be determined by a variety of factors. The location itself will determine whether bats are present all the year round, or just at particular periods in the year. This implies that several surveys will be needed over the year. For example, a study of the greater noctule would require surveys during the late summer migration period, even in areas where there is no summer population. Locations where food is available only for short periods should be surveyed when there is a plentiful food supply.

As a rule, not all species will be recorded in a single night. A survey might succeed in detecting all the species in a location where there are only a few species, and where these species are common. However, small daily changes in the activity of those species present in the habitat will necessitate repeated surveys on at least two or three consecutive nights. Even then, it is unusual to record all the species present.

The diagrams in Figures 2.2, 2.3, 2.4 and 2.5 show examples of how the range of species develops in the course of a year, and how the length of recording and the number of repetitions affect the results. The first two figures come from a long-term survey recording at 10 m above ground level and about 25 m from a row of trees or forest edge.

Over the course of the year the number of species documented in one night increases, to decrease again at the end of the activity period (Figure 2.2). The maximum number of species recorded at this site on a single night was eight, and this happened on only two occasions. Nights with seven species were more frequent. Generally, however, only three species were recorded, as can be seen clearly from the second diagram (Figure 2.3). Many nights with only one or no species also featured. This is always to be expected, particularly on nights of heavy rain.

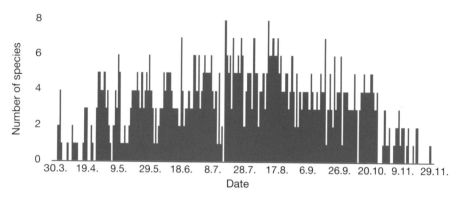

Figure 2.2   Number of species recorded in a continuous survey (open area at a height of 10 m and a distance of about 25 m from a tree line or woodland edge).

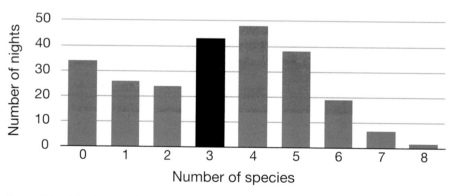

Figure 2.3   Distribution of the number of species each night from the raw data shown in Figure 2.2. The median value is highlighted in black. The mean value is around 3.1 species per night.

The maximum number of 12 species at this site was only reached after a long survey period of over 170 days. In order to illustrate this issue with other data, five further studies with 10 to 13 species on sites ranging from open ground to housing areas were evaluated in the same way (Figure 2.4). In this graph, the survey described above is represented by the orange curve which reaches the 90% threshold on day 137. The twelfth species was a Bechstein's bat, which is usually a difficult species to identify from acoustic data, and one which was not expected at all at this location. It is also evident that the survey nights early in the year, with few species each day, do not yet give a representative picture. If the survey had started four weeks later in May, for example, the accumulation of species would have occurred much more rapidly. Species 9 to 12 were only infrequently picked up, and so would presumably have been overlooked. In this example, however, the location had no particular significance for these species, so this could explain their infrequent occurrence.

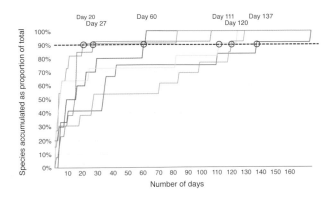

Figure 2.4 Species accumulation over time at six different locations (shown in six different colours) with 10–13 species. The red dotted line shows the 90% threshold which is reached after an average of 80 days (range 20–137 days).

This type of build-up in the number of species is often due to infrequent visits by individual rare species. Seasonal activity patterns such as the breakup of the maternity roost or migration may also impact strongly on the range of species at a particular location. The animals then become more mobile and are encountered over a wider area as they disperse.

Surveys in a natural woodland reserve in northern Bavaria in Germany clearly illustrated the need for repeated recordings to measure diversity (Figure 2.5). In this example, data were collected in parallel in each of three locations over two whole nights a month during the summer. In the course of the recording, new species were added in each month. In one potential roost site (S2) this figure was 13 species. This actually represents the maximum in this area. In other locations species were absent, so that maxima of 3 to 10 species were recorded (S1 and S3). The figures in the following year turned out to be significantly higher. This is the case for location S1 in the example. To summarise, various factors, such as season, year and weather, all can have an impact on the range of species.

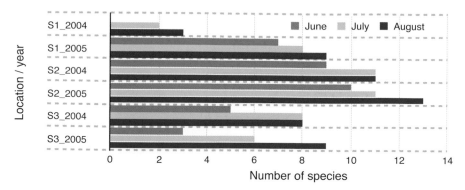

Figure 2.5   Number of species recorded during surveys at three different locations. For any one location, the survey was completed under comparable conditions on each occasion.

3 The planning of acoustic studies

The goal of acoustic recording is the investigation of bat activity within a specified area. The area is usually defined by a scientific study or by the type of intervention/development that is being evaluated. The fundamental questions that need to be asked will determine the depth of the investigation and the approach taken. In terms of practical implementation, it is necessary to decide precisely how the technology is to be applied in order to obtain the required data.

In Chapters 1 and 2 we set out the basic principles of acoustic recording, and surveyed some of its typical applications. In this chapter, a number of scenarios are illustrated by means of simple examples. Both active and passive approaches are introduced. The main objectives of these projects are a good coverage of the area and, ideally, the identification and description of important habitats and partial habitats.

In planning a project, it will often be necessary to take account of existing guidelines or requirements for feasibility studies, such as those for the initial planning of wind turbines, and also their subsequent operation. Similarly, there may be stipulations for road building, particularly major roads and motorways, which are usually agreed at central government level. There are mapping instructions or recommendations for single species groups and evaluation, for example, within the framework of special conservation measures for species. While guidelines are binding in principle, deviations are always possible if there is a sound scientific justification. Such deviations may be important in order to respond flexibly and effectively to the requirements of the project, and must take account of parameters such as the range of species, the habitat or the individual nature of the intervention. In some circumstances, it may be necessary to work around the weaknesses of a guideline. In this situation it is essential to have not only good knowledge of bat ecology but also a good understanding of acoustic methodology. The inappropriate application of the methodology may so impair the recording that the results will no longer be scientifically robust.

Reliable results are always achievable with large datasets obtained from quantitative recording. In the case of conservation planning for partial habitats, qualitative recording and the documentation of behaviour, such as courtship displays and flight paths, will be necessary. In order to describe the activity, it is helpful to ensure as complete a coverage of the area as possible, as well as comprehensive sampling. The recording must be as objective as possible, and the cost of the study must not exceed the available budget.

3.1 Documentation of activity

Once the activity has been recorded, it must be interpreted in order to generate usable data. In this section we illustrate some possible key data for a comparative survey of the activity recorded in an area. Further key data, such as the number of bat passes, the number of recordings, sum of seconds with activity or time categories with activity, are presented in detail in a later chapter (Section 10.2, *Quantification of activity*). In this chapter they are only described briefly in the context of recording methods. The total number of bat passes recorded in a session as a measure of activity is inadequate, since activity, as will be seen later, is very difficult to define. Relative abundance can be determined by the number of recordings per unit of time, but this is still not a fail-safe method, as it is strongly affected by the behaviour of the bats. Time intervals with activity are also a measure of abundance. Both of these methods are only feasible when using automated systems.

Dot density maps can be created by entering a dot on a map for each location where a bat was recorded. It then becomes clear where the hotspots of bat activity are in an area. This measure of relative abundance can only be achieved by the bat worker moving around the area with a hand-held device. Static devices are of no use in this situation. If the bat worker walks along the transects at a standard even pace, the physical distances between the bats recorded can be used as a measure of abundance. It is vital that no bats are missed during the walk along the transect, so the parallel use of an automatic recording system is absolutely essential.

If the statistics for relative abundance are measured in units of distance or time, then an average relative abundance for the whole area can be determined. Spatial sampling from transects or fixed locations can also be drawn on. It is also possible to establish median values for all partial habitats and relate them to the total area of the study. In this way, a figure for the relative population size in a habitat or study area can be derived by multiplying the relative abundance by the area of each partial habitat. However, a large number of samples will be required for the calculation of the relative carrying capacity of a habitat.

Activity may also be described qualitatively, when there is documentation of such aspects as the type of activity (such as hunting or social calls), consistency of behaviour (long or short term) and the number of bats.

3.2 Definition of partial habitats

In preparation for some of the approaches to surveys described in this chapter, it will be necessary to investigate each of the partial habitats within the survey area for their use by bats (Figure 3.1). This often turns out to be far more difficult than it initially appears. The categorisation of habitat types relies, firstly, on the precise nature of the study, and secondly, on the actual structure of the landscape and possibly the range of species present. The procedures used in a survey carried out in primarily open landscape will be very different from those appropriate to a woodland area.

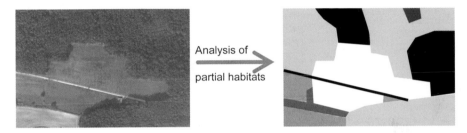

Analysis of
partial habitats

Figure 3.1   In the preparatory phase of acoustic recordings, it is necessary to carry out an assessment of the partial habitats in the area. Surveys such as those described in this chapter can then be planned.

There are various approaches to describing and categorising habitats, and these are frequently based on natural vegetation or biotopes. Most systems for describing habitats use very precise categories, but the use of these categories with bats is questionable. As a rule, particularly detailed habitat classification is not necessary with animals that often change habitats. Such animals typically move around opportunistically when searching for food. Even if single individuals have favourite foraging areas, these preferences cannot be fully explained either by the structure of the vegetation or by the density of insects present. These aspects should therefore be considered when defining partial habitats.

For example, a hedge in an otherwise open landscape serves not only as a potential hunting ground, but it also links possible partial habitats and will have an important role as a well-established commuting path. In open landscapes dominated by agriculture, wildflower strips may also be important for bats. It will, however, be difficult in the preliminary study to establish this convincingly and comprehensively. In woodland, on the other hand, there are likely to be numerous potentially significant structures available in the areas in which bats can fly. In some circumstances, the precise mix of tree species is less important than the structure of the habitat. If the area of forest is considerable and generally uniform in terms of management, it can be easily researched by means of forestry records or even by the use of aerial photos. However, important structures may be hidden within the stands of trees. Figure 3.2 is a suggestion for a schematic representation of the structures in woodland habitat.

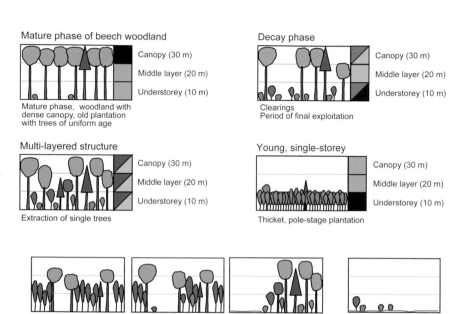

Figure 3.2   Examples of woodland structures which are of importance to bats (adapted from Runkel 2008). The distinctions are based on the density of vegetation in the various layers. Bats develop flight paths within these layers. The examples here are a mature beech woodland with a dense canopy (top left), a multilayered woodland (centre left), a woodland that is in a state of decay (top right) and a young plantation (centre right). Various linear woodland structures and open areas with trees are depicted in the bottom row.

3.3 Systematic sampling

Once a comprehensive habitat analysis has been completed, data can be collected using both stationary and mobile techniques, so that all the significant habitat types are covered. The survey must include potential roost sites, flight paths and high-quality foraging areas in order to cover important functional components and document the full range of species in the study area.

3.3.1 Systematic stationary surveys

For the purposes of static, passive recording, individual locations are specially selected (Figure 3.3). These locations should be at potentially significant points, but they do not necessarily have to be numerous. If a sound representative sample is selected, a few good locations are adequate, and will provide useful phenological information. By means of appropriate analysis techniques, using units of time, for example, statistics on relative abundance can be obtained. If the size of the partial habitats relative to the whole area is calculated, average relative abundance figures can be derived.

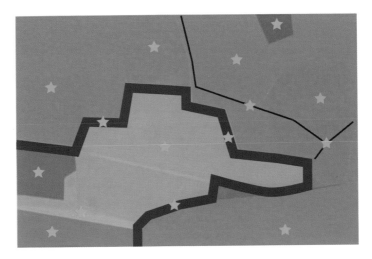

Figure 3.3 Once a partial habitat analysis has been carried out, locations for passive recording can be systematically chosen.

3.3.2 Systematic mobile surveys at fixed points

Regularly distributed stopping points for recording are defined along carefully planned transects. These points are situated in potential partial habitats (Figure 3.4) that are mainly located in the heart of the study area. By employing this method, relative abundance and descriptive assessments of the activity at the fixed stopping points can be undertaken.

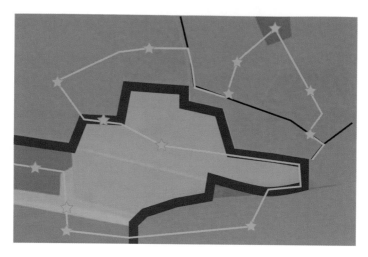

Figure 3.4 After a partial habitat analysis, transects with fixed stopping points for active recording are carefully selected.

In comparison to transect walks carried out at an even pace, more precise activity data are obtained at the stopping points. Longer observations at the predetermined points allow a more thorough investigation of the role of the partial habitats. The disadvantage of this system is that activity data are only obtained for the stopping points. It is therefore important to have enough stopping points to ensure that the data are representative of the whole area. The period when the observer is moving from one stopping point to the next represents dead time as no recording is being carried out. In the end, just as is the case with systematic stationary recording (Section 3.3.1), the result may be strongly influenced by the subjective selection of the stopping points.

3.3.3 Systematic mobile surveys with uniform coverage

Before a survey begins, fixed transect routes are established. The observer walks the transect at a constant even speed several times during the course of the year. To ensure a representative data sample, the transects must pass through all the important partial habitats (Figure 3.5), with the main focus on the core study area. In this way, statistics will be obtained for the average relative abundance for the whole area and for the relative abundance for individual transect sections. Dot density maps can be drawn up, and hotspots of activity identified.

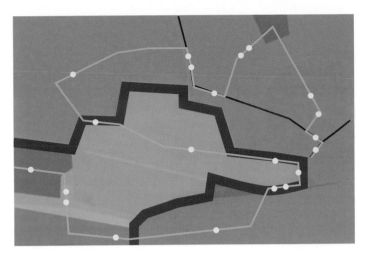

Figure 3.5 After analysis of the partial habitats, transects are carefully selected and monitored at an even walking pace.

The process is notable for its good reproducibility and its high degree of objectivity, particularly when compared to a survey using stopping points previously selected by the observer. Even inexperienced bat workers can record bats along the established transect routes if an automatic detector is used. No particular skills or information are required, as they are when using pre-selected stopping points. Uninterrupted listening for bats throughout the transect will provide significantly more data, even if there is a danger of replication of recording individual bats, so-called pseudo-replication. If transects are confined to a limited number of partial habitats, statistics can be obtained for these areas alone.

Fixed route planning can have a negative effect on the coverage of the area. It is not always possible to find appropriate paths through all partial habitats. Rough terrain, in particular, makes it difficult to maintain a steady, even pace, so that, at the planning stage, it might well be that such areas are totally or partially ignored. The survey will then tend to be more superficial, as it is not possible to stop, even briefly, to carry out qualitative observations in more promising areas where there might be foraging, courtship behaviour or roosts. Finally, it is a matter of chance whether bats are present as the observer passes.

3.3.4 Systematic mobile freestyle surveys

In the lead-up to a survey, important partial habitats are likewise identified and integrated into the planning. The transects are then monitored, but without prescribing the pace of the walk. The approach is more *freestyle* as unplanned detours or other changes to the route are also permitted (Figure 3.6). While the transects are concentrated in the core study area, the intention should be to cover the whole area. Longer observations should also be planned at activity hotspots, to better document qualitative aspects. On the basis of the resulting data, the activity can be assessed in terms not only of species, but also of individuals and behaviour. At a pinch, the statistics for relative abundance could be presented in the form of dot density maps.

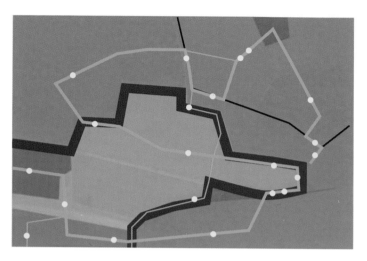

Figure 3.6   After a partial habitat analysis, transects for active recording are carefully selected. These can be adapted and modified when the transect is walked.

Because the bat worker can adapt the transect in the field or make detours, all the important partial habitats can be covered and qualitatively assessed, as the stopping times can be altered when appropriate. Furthermore, interesting areas can be monitored at the optimal times, since there are no hard and fast rules for the walking speed or for stops. This is particularly helpful when investigating roosts or flight paths. This method is also especially valuable in difficult terrain.

The advantages, however, come at a cost. Because of the changing and variable stopping times, each partial habitat will have fewer data collected for it as the year progresses. This is a sampling artefact, as there will be a tendency to reduce the overall time spent on the investigation of any one habitat. Because of this variability in the stopping times and locations, the dot density maps are less meaningful and there are no quantitative activity figures. This method is primarily suited to experienced bat workers, because the evaluation of the activity and the decisions that have to be made in the field are subjective.

3.4 Random sampling

In methods which are based on random sampling, it is possible to dispense with detailed habitat analysis before starting the survey. The surveys are carried out in randomly selected locations or along randomly selected transects.

3.4.1 Random stationary surveys

The surveying takes place at randomly chosen locations, which obviates the need to carry out a prior habitat analysis (Figure 3.7). With the resulting data from a few single-night surveys, useful information relating to phenology at night can be obtained. If the location is studied over a longer period, more detailed conclusions can be drawn about the phenology. Furthermore, statistics on relative abundance as well as average relative abundance can be obtained for the whole survey area. Since there is no subjective pre-selection of locations, the process is absolutely objective.

Figure 3.7   A large amount of passive monitoring must be carried out at locations chosen at random

As a large number of locations are usually surveyed, so that representative data are obtained and no important partial habitats are missed, the effort required is enormous, both to carry out the surveys and to analyse the sheer amount of resulting data. Finally, the time patterns only allow a tentative conclusion to be drawn about the roles of the partial habitats.

3.4.2 Random mobile surveys

Transects are placed randomly through the survey area in such a way that the whole area is more or less fully covered (Figure 3.8). Again, no preliminary analysis of partial habitats is necessary. The surveying of the transects has no fixed methodology, so techniques using stopping points or walking at a constant pace are all equally valid. The survey area will be well covered with a continuous uninterrupted sweep along the transect. The greatest advantage is the objectivity of the survey, since there is no need to make a subjective choice of partial habitats.

Figure 3.8   Random transects through the area with a free choice of methods.

Conclusions regarding phenology are not necessarily reliable. This may potentially have a negative effect on the data for the range of species, and it is only with considerable effort that reliable data will be obtained. Equally, the statistics on abundance are of only limited validity.

4 Manual and automatic acoustic recording

Recording of activity, using manual and active or automatic and passive data capture, has been mentioned in the preceding chapters. There are fundamental differences between automatic and manual acoustic detection of bats which are evident in aspects such as subjectivity, reproducibility and quantity of data. The differences affect the applicability of these methods, the results of surveys carried out and the subsequent interpretation of the data. Comparisons of the results of the two methods may be possible, since they both have acoustic activity as a basis for their data collection, but comparability of data is not a given. The differences between the methods therefore have to be taken into consideration during analysis. Acoustic data from various sources must not be collated uncritically and compared directly.

It is generally the case that the results of acoustic recording originating from different devices cannot simply be regarded as equivalent. The modes of recording strongly influence the results obtained, and differences between datasets are strongly affected by their application. Mostly, however, the range of species and any activity index of species derived from it are comparable. To what extent this is meaningful, however, depends essentially on the exact protocol of the survey.

4.1 The principles of manual recording

Manual bat detectors make ultrasound audible to the human ear. These are usually heterodyne or frequency division, which are described in Sections 14.3, *Heterodyne detectors*, and 14.4, *Frequency division detectors*, respectively – although occasionally time expansion detectors are used by more skilled surveyors. A bat worker is able to hear bat calls directly with the detector and can adjust the frequency (with a heterodyne detector), the volume and sensitivity of the device. Calls are recognised simply by listening. Thanks to the functioning of the ear and the processing of the resulting stimuli to the brain, humans can perceive very low-volume calls or even the sounds of rustling or static. This makes the human operator very suited to the investigation of bats using detectors. This is particularly true when general bat activity is being surveyed over a period of a few hours, and when there is no need for precise identification. This could be the case, for example, when carrying out preliminary surveys to find locations which will be investigated more thoroughly later. Thanks to the mobility of the recorder, large areas can also be covered in a short time.

However, a human operator cannot work at locations such as wind turbine nacelles, wind and meteorological masts, or similar sites. Quantitative and qualitative recordings for comparison between locations or different nights are similarly not without their problems. Furthermore, a human operator cannot carry out long-term recording, as even two or more hours of intensive detector work is very exhausting. Long periods of manual recording will thus have a negative effect on the quality of the results, because concentration declines and bat calls that are low volume or of short duration will be more easily missed.

4.2 The principles of automatic recording

Automatic recording is based on the recognition of acoustic calls in real time by the analysis of certain signal properties such as volume or frequency. In addition, parameters are analysed, such as a trigger threshold (amplitude), frequency ranges contained in the signal, and signal properties which can be derived (amplitude flanks, energy share in the spectrum). As a rule, all signals which are above the threshold level will trigger the system and lead to further analysis. In this way, a recording is only triggered by calls which exceed certain threshold criteria set by the user. This allows a reduction in the number of false-positive recordings which would otherwise arise from only monitoring call amplitudes. To what extent this works reliably, and to what extent bats are missed, depends on a number of factors, including the precise settings.

Through an analysis of the signals and the resulting trigger settings, an objective determination of bat activity can be achieved. An echolocation call of identical shape will always trigger a recording. The algorithm must operate conservatively, so that even unclear calls will trigger the system. Similarly, calls which are emitted against the background of other noise should also be recorded, provided that these calls are not totally drowned out.

It will always be difficult to achieve a workable system, and there will be the constant risk of missing echolocation calls. It will also be difficult to fine-tune the system so that it will not be too sensitive and produce too many false-positive recordings. Locations with significant background noise are problematic, especially when there is an ultrasound component, such as from nearby train or tram lines, roads and wind turbines. Bush-crickets in summer may also result in many wasted recordings.

In the above situation, it is helpful to filter out at least the lower-frequency sounds, since, for example, a large proportion of bush-cricket calls in central

Europe are under 18 kHz. Some of these unwanted signals can be excluded by means of a suitable high-pass filter. The lower the frequency that the high-pass filter is set to, the greater will be the number of signals that trigger a recording. A higher frequency setting of the high-pass filter, on the other hand, will filter out not only the low-frequency bat calls of such species as European free-tailed bat and greater noctule, which are regularly under 14 kHz, but also the social calls of a number of bat species. Unfortunately, the carrier frequency of some common, loud and continually singing bush-crickets (Tettigonioidea) is also in the same range.

The differences between the various devices will ultimately have an influence on the data obtained. The trigger threshold for recording, the amount of downtime and the duration of recording can be decisive factors in the subsequent analysis. Not every system, therefore, is equally good for autonomous operation, and the one chosen will depend on the technical requirements. For example, the lack of a power supply in the field when carrying out long-term monitoring will rule out some systems. Often it is not possible to carry out a calibration or test the system in action, which is particularly important for long-term monitoring. The consequence of this is that microphone damage and failure are often only spotted after the completion of the survey.

In Chapter 9, *Criteria for detector systems*, there is a discussion of the optimal detector system and of the aspects which affect the data.

4.3 Comparison of manual and automated recording

The great weakness of human bat workers is that they find it difficult to maintain concentration constantly for surveys over one or more nights. Subjective errors will always creep in, and mistakes by any one individual could potentially compromise the comparison of the results of several bat workers. The use of certain types of detector technology can indeed mean the exclusion of whole species groups if the setting is incorrect, for example an incorrect frequency choice on a heterodyne detector. It is, furthermore, impossible to quantify subjective errors, which might allow the retrospective correction of the data. Nevertheless, the human bat worker can often estimate activity in the field quickly and accurately on the basis of visual observation of the bats that have been detected. Sufficient experience is really what is needed here.

Automatic broadband systems have many advantages, above all for long-term recording and comparability of data. The whole frequency range is monitored, and the sensitivity of the device is always identical, as long as the settings have not been changed by the user. This is vital if comparable data are to be obtained.

The disadvantage of automated recording is the generally lower distance range, since a fixed amplitude threshold has to be exceeded in order to trigger a recording. The threshold must not be set too low, as significant noise disturbance can lead to a high proportion of unnecessary, wasted recordings. This is important so that meaningful automated call recognition can be carried out and storage space is not taken up unnecessarily. In the light of these issues, it should therefore never be forgotten that the human ear can detect even the most indistinct of calls!

In order to pick up bat calls over longer distances, the sensitivity of the call analysis or the amplification of the sound signal can be increased. In many cases, however, the number of usable calls recorded barely improves, because of increased hiss from the devices – the so-called signal-to-noise ratio remains unchanged. Many of the low-volume calls are thus no longer identifiable, since crucial components of the calls are lost. As a result, automated recording is made considerably more difficult. Nevertheless, if the distance range is set fairly low, the automated systems prove to be indispensable tools for the recording of activity. By reducing the distance range setting, data can be collected, for example, in small, confined areas.

Table 4.1 provides an overview of the comparative performance of manual and automated detection of bat activity.

The laws of physics describe the limitations of the spread of ultrasound, and thus the maximum detection range. Calls of 20 kHz can at best be heard at distances of over 100 m, whereas the corresponding figure for signals of around 40 kHz would be about 30 m. In general, higher frequencies fade more quickly than lower ones (see also Section 14.1.3, *Propagation of sound*), an aspect which must be borne in mind when planning acoustic surveys. This is dealt with in more detail in Section 9.3, *Recording distance and amplitude*. The general rule of thumb is that, as the frequency increases, detection becomes more difficult. Bats in the distance, whose calls have been recorded only faintly, are often impossible to identify accurately. They increase the size of the dataset, but rarely provide better results.

Moreover, the bats often do after all approach closer to the microphone, triggering a recording, but one which is of a shorter overall duration. For this reason, it is possibly better to carry out the data analysis without the inclusion of bats in the distance. The generally lower distance range of automated systems is compensated for by the objective triggering system and the genuinely long-term recording which is possible.

The different methods of bat recording do not all have the same sensitivity for all species. Other forms of ecological data collection suffer from this shortcoming too, and significant conclusions can nevertheless be drawn using these data. A certain selectivity must be accepted and allowed for in the interpretation of the results.

The selectivity of the heterodyne detectors, which is dictated by the limited frequency window, does present a problem. Above all, those species with high-frequency calls, such as soprano pipistrelle or horseshoes, will

Table 4.1 Comparison of manual and automatic recording

	Manual recording (heterodyne detector)	Manual recording (frequency division detector)	Automatic recording (zero crossing system)	Automatic recording (direct sampling system)
General range/sensitivity	High	High/medium	High/medium	Medium
Selectivity (species)	Medium	Low	Low	Low
Mobile recording	Yes	Yes	Yes	Yes
Recording duration	Short, maximum of a few hours	Short, maximum of a few hours	Very long, up to several months	Long, up to a few months
Species recognition	In the field	In the field and limited computer analysis	Slightly limited computer analysis	Computer analysis
Reproducibility/comparability	Low	Medium	High	High
Costs	Staff, recurrent	Staff, recurrent	Device, one-off	Device, one-off
Multitasking	No	No	Yes	Yes

be missed more easily when the detector is not tuned to higher frequencies and thus will not be included in the species inventory. This can be resolved by using frequency division detectors, which convert the whole ultrasound range into signals audible to the user (broadband). The drawback is that species recognition in the field is significantly more difficult with these devices. They are also more subject to disturbance from grasshoppers than the heterodyne devices. Devices for automatic recording are, as a rule, all broadband, which is a great advantage if non-selective detection is required. However, some triggering functions, which control the recording, do not exhibit the same sensitivity for all frequencies or call types. This of course means that there will still be a certain degree of selectivity.

The human concentration span limits the use of a bat detector for immediate species identification in the field to between two and four hours. Moreover, the operating time of smart phones and tablet-based detector systems is limited by their battery life, and this also restricts the length of a recording session.

The power supply of automatic systems is similarly a limiting factor, although traditional automatic recording devices do in general operate for significantly more than four hours. Good systems will last up to a week. For long-term recording over several weeks with automatic systems, it is usually necessary to use an external long-life battery (e.g. lead acid) or an alternative power supply, such as solar panels, to charge the batteries. The latter will allow autonomous operation over long periods of time, provided that there is sufficient solar radiation. Devices which are linked to a computer are only suitable for long-term use if they are connected to a mains supply. This would obviously rule out their use in some locations.

The identification of bat species in the field depends on hearing and is thus very subjective. By using time expansion detectors, however, the identification can be checked on the computer (this is also the case with frequency division detectors although call analysis is more challenging due the reduced clarity of the sonograms). Time expansion detectors buffer 1–3-second digital sound fragments at high resolution. These can then be slowed down by a factor of 10 and transferred onto a storage medium. This process generally takes between 30 and 120 seconds, and during this time no other bats can be detected. A further disadvantage is that very low-volume calls can often also be heard and are recorded by the user, or the user makes noises when walking. These junk sounds can make species recognition more difficult. In order to obtain reliable results, the recorded calls must be of sufficient quality. Low-volume calls, calls with hiss, or those that are fragmented, are only rarely of use. While this can be a particular problem when automatic call analysis is carried out, this is the only way to process large quantities of recordings.

When relying on manual recording, the comparability of data is improved if it is always the same person conducting the survey. This does mean that only sequential comparisons can be carried out, as it is not possible to do simultaneous surveys at different sites. With species that are simple to identify, such as the common pipistrelle, the data from various people can be compared provided that the same survey protocol is used.

With most surveys, a large number of outings in the field are necessary in order to obtain robust data. The staff costs for manual detection will thus be correspondingly high. With automatic recording, in contrast, the initial investment in equipment will be high, but it will be more cost-effective in the long run because of lower staffing requirements and, usually, significantly better data quality.

While automatic recording is preferable, that does not make manual methods obsolete. For many surveys, field work will always be necessary for at least some of the time. People are needed for interactive tasks, such as finding roost sites or investigating flight paths. Manual recording can be carried out over the whole of the survey site, whereas automatic data collection is often restricted to only a few sites.

5 Manual identification of species

One of the aims of acoustic recording is the identification of bat species by the analysis of calls. Similar to bird song, bats have developed species-specific 'social calls' to advertise their presence. In contrast, the echolocation calls of bats do not serve to promote their species (Barclay 1999). They tend to be used for navigation and finding prey, and are adapted to a task that has been determined by their environment (Jones 1995; Schnitzler and Kalko 2001). For this reason, the echolocation calls of some species are very similar or even overlap completely. Identification may therefore be tricky, and, in some cases, it may not be possible narrow down the identification to a precise species (Obrist *et al.* 2004).

Manual identification may be performed either by listening directly in the field with a heterodyne, frequency division or time expansion device, or by listening to and analysing recordings on the computer (Zingg 1990; Jones *et al.* 2000). To complement manual identification, there is of course automatic recognition of recordings, for which there is now a good supply of software tools (Fritsch and Bruckner 2014). This is dealt with in the Chapter 6.

5.1 Manual identification in the field

In manual recognition, the bat worker identifies the species simply by listening to the sounds heard directly on the bat detector. Differences in the processes used for converting ultrasound between, say, heterodyne and frequency division devices will also affect the sound produced and the identification of the individual species. The specific device used may also exhibit similar variations.

5.1.1 Identification by ear

This technique used to be the only way of detecting and identifying flying bats. The detector simply converts bat calls into sounds that are audible to humans, allowing the user to identify the species. With the frequency division detector, pitch and rhythm are used to aid identification. In contrast, the user of the heterodyne device can select the frequency and use the rhythm and the impression of the sound (wet or dry) to make an identification. With time expansion systems, skilled surveyors can recognise call length, where intensities of signal strength occur within the calls and other characteristics.

So, for example, the calls of bats of the genera *Myotis* and *Plecotus* sound dry, whereas those of *Pipistrellus*, *Eptesicus* and *Nyctalus* sound wet (see Section 14.3, *Heterodyne detectors*). While the common pipistrelle has a somewhat irregular rhythm, in the UK the Nathusius' sounds more regular than its congener. The serotine bat, in contrast, has a regular rhythm with a high repetition rate, but there are frequent breaks in the call sequence, creating a syncopated effect. Heterodyne and frequency division detectors are both very effective tools for identifying numerous species of bat in the field. With practice, certain groups of species can be identified fairly confidently. Identification is made easier when the call can be heard for a longer period of time, because the bat is then more likely to emit its characteristic species-specific sounds.

There are objective descriptions of typical call characteristics of species that can be characterised easily, for example the 'chip-chop' of the noctule, or the 'quail calls' of the pond bat. Moreover, many bat workers also have their own, subjective criteria which are the result of experience gained over time with the bat detector. Observing a roost where the species is already known is one way of improving identification skills with the bat detector and becoming familiar with calls made by particular species. Nevertheless, care has to be taken that there are no other species in the vicinity. A much more reliable way is to record the bats using a time expansion detector or a direct sampling device and analyse the calls later at one's leisure. There are now many software programmes which can simulate the sounds of a heterodyne detector, allowing the user to recapture that 'in the field' experience.

5.1.2 Visual clues

Bats can often be more reliably identified if the sound technology is supported by direct visual observation. Identification is made much easier if features such as size, silhouette or flight style can be noted. Good observation is, of course, only possible if light conditions are suitable, such as at dusk or dawn. The use of lighting in absolute darkness is not usually effective, as many species are disturbed by direct light and will take evasive action. When the bat is only seen briefly, visual confirmation of identification is not usually possible.

5.1.3 Advantages of identification in the field

Working with a bat detector in the field has many advantages over automatic, passive systems, in addition to the mobility of the bat worker. The behaviour of the animals can be observed directly, making it possible to draw qualified conclusions about activity. With practice, different types of behaviour can be distinguished, such as whether bats are feeding or are simply moving along their established flight paths. Swarming behaviour is particularly easy to recognise, for example, and can help in finding roost sites at dusk and especially at dawn. People with particularly good hearing will be able to identify bats quickly and confidently once they have acquired a good knowledge of the sounds that bats make.

5.1.4 Disadvantages of identification in the field

When using a bat detector in the field, there will at times be some doubt over the accuracy of the identification, because of the element of subjectivity. While some bats, such as the common pipistrelle or noctule species, can be confidently identified, even by inexperienced field workers, others such as the *Myotis* bats pose considerable difficulties, even for experts. Unless a recording is made, it is impossible to verify the identification independently, or for the bat worker to check their own assessment. Identification from memory is, moreover, always prone to error. All these factors may easily lead to incorrect data.

A bat worker will always benefit from experience, but it is not always the case that an experienced person is better or more accurate in their identification. In practice, there will always be situations in which correct identification is difficult. The calls of native bat species overlap, both in rhythm and in frequency range, so that brief and rather atypical encounters will tend to produce higher error rates. At the beginning of the season, field workers should always reacquaint themselves with all the different bat calls.

5.2 Manual identification of recordings

Manual species identification may also be carried out using computer-aided analysis of sound files which have been recorded using time expansion detectors or direct sampling passive systems. This technique involves the graphical presentation of digital recordings and allows greater precision. Oscillograms, for example, show amplitude plotted against time as well as call pulses, call lengths and call intervals. Spectrograms and sonograms, on the other hand, illustrate frequency distribution. The spectral analysis of individual calls is used for the extraction of frequency parameters. A more detailed discussion of this can be found in Section 14.7.1, *Representation of wave form*.

There are a large number of software programs available in this field, including Avisoft, BatSound, Audacity, bcAnalyze and many others. Although these all generally work according to the same principles and use the same algorithms (primarily FFT), they can produce widely differing outcomes. The reason for this is the large variety of settings and internal processing methods used in the FTT calculations such as window type, window size and overlapping. These variations may even be so great that the identification yields different results.

5.2.1 The process of identifying species

The process of identifying species is generally based on both an evaluation of the sequence of calls and a precise consideration of the frequency patterns shown on sonograms (Figure 5.1). Oscillograms, for their part, show rhythm and call intervals, and the number of individuals recorded. To identify the species, individual calls in the recording are shown in the form of a sonogram that portrays frequency against time for a selected sample of the recording. The shape of the call and its frequencies are used as criteria for identification.

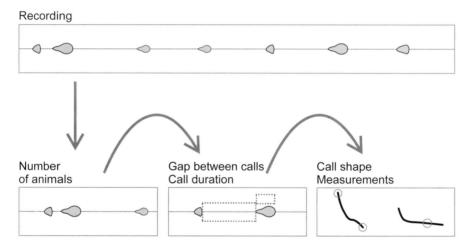

Figure 5.1   There are several steps in the manual recognition of recordings.

A distinction is made between (quasi-)constant-frequency calls (CF/QCF), frequency-modulated calls (FM), and those which have both an FM and a QCF component (FM-QCF) (Figure 5.2). The call shape represents in itself a very important criterion for distinguishing between the genera. CF call components are used by the Rhinolophidae, while QCF calls are mostly employed by *Nyctalus* and *Vespertilio*. The calls of the latter two genera cross over into FM-QCF calls, which are also used by *Eptesicus*, *Pipistrellus*, *Miniopterus* and *Hypsugo* (Figure 5.3). *Plecotus* and *Myotis* emit FM calls.

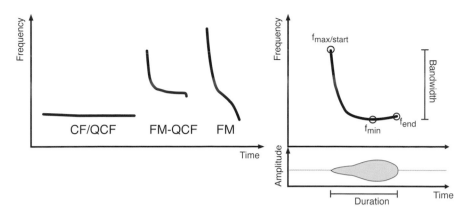

Figure 5.2 On the left are the sonograms of various calls, categorised by call shape, which is determined by the frequency plotted against time. On the right, typical sonogram parameters are illustrated.

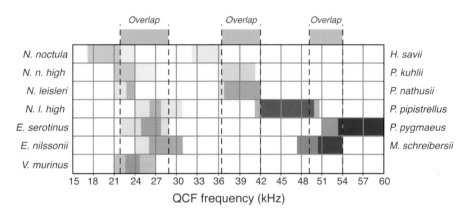

Figure 5.3 The main frequency ranges of various species with QCF calls. Ranges in which there is considerable overlap are highlighted.

The call shape and the frequency spectrum are the main parameters used to identify species, but some authors also use the amplitude wave. The characteristics of the call are portrayed on a sonogram, which may also include the start and end frequencies and any kinks or bends in the call shape. There is, however, no generally valid definition of what parameters are included, and some variation is possible (see Section 8.4, *The impact of the analysis*). It is indeed not unusual for authors of reference works to give no information at all on the parameters used.

In Chapter 8, *The complexities of call analysis*, there is a more detailed consideration of possible errors which arise in computer-aided call analysis. The following analysis concentrates more on technical concerns than on the impact of human error and other general factors.

5.2.2 Limitations of the process

Reliable identification is not always possible if a recording only includes single calls, but the task becomes progressively easier as the number of calls increases. Recorded calls do, however, need to be of sufficient quality, as identification becomes harder or even impossible if any parts of the call are missing or are not measurable. The use of time expansion detectors is particularly problematic in this respect, as they generally produce few viable calls, and the recordings are subject to a deterioration in quality during the multiple analogue-to-digital/digital-to-analogue conversions.

Even when good recordings are available, there is a lack of the sort of information that can be gleaned from visual observations in the field, whether short or long term. It will therefore be necessary to make decisions on the basis of short call sequences.

5.2.3 Advantages of manual identification of recordings

If recordings are stored on the computer, it will be possible for them to be cross-checked later by other bat workers. Recordings can, furthermore, be of great value for spreading newly acquired knowledge to other field workers. None of this is possible with simple active detection and identification in the field.

5.2.4 Disadvantages of manual identification of recordings

Non-automated identification of recordings is prone to the same subjective errors as identification in the field. Scientific research shows that the level of experience of the person analysing the data has a significant impact on the quality of the identification. Since there are no definitive reference libraries of bat calls, and since some species of bat have such variable calls, the acquisition of the necessary experience is a very time consuming and involved process. Learning to recognise typical calls of any species is liable to error because of the difficulty in obtaining reliably identified recordings of species in the field which can be used as benchmarks.

Identifying bats without the help of automated systems is a very slow and painstaking process, and the analysis of one recording can take between 30 seconds for the simpler species and several minutes for the more difficult ones.

While it would be nice to narrow down every recording to the exact species, this is simply not possible if the recording is insufficiently clear. It will result in a great deal of wasted time and will very likely lead to subjective errors. Subjectivity can mean that there are sometimes considerable discrepancies in identification between different people. Nevertheless, it should be possible to have identifications checked by an independent third party. In the context of environmental impact assessments, for example, recordings of rare species are sometimes sent to experts for confirmation.

The problems of automated call recognition caused by the overlap between species, as well as the reference calls of the difference species, are also issues in this context, and are dealt with in Section 6.1.3, *Limitations of the process.*

6 Automatic species recognition

Automated identification of bat species has increased enormously in importance since the introduction of automatic recording systems. Its usefulness cannot be underestimated, particularly in the processing of very large quantities of data. However, because bat calls vary considerably, even within species, these automated systems do not always give reliable results. Consequently, this remains a controversial area, especially in the world of specialist scientific research (Russo and Voigt 2016).

This chapter gives an overview of the process of automatic species recognition, without, however, going into detail about the various individual systems. Finally, there is a discussion of the opportunities opened up by such systems, as well as their limitations.

6.1 Automatic identification of recordings

The beginnings of automatic identification of bat recordings can be traced back to the middle of the first decade of this century. The first simple systems were developed in Britain and Switzerland for research projects (S. Parsons and M. Obrist). The first commercial system, available in 2007, was ground-breaking and included a comprehensive species list (bcDiscriminator, ecoObs GmbH). Since then, new products have become available, including bcDiscriminator's successor, batIdent, as well as other systems for European species from Great Britain (iBatsID), Switzerland (Bosch and Obrist 2013), France (SonoChiro, Biotope) and the USA (Kaleidoscope, Wildlife Acoustics). Systems have also been developed for the fauna of other biogeographic regions.

The increased use of passive monitoring systems, some of which are in constant operation, has necessitated the processing of ever greater quantities of data. Automatic recognition systems have consequently enjoyed a boost in popularity. Common to all systems is that they automatically extract and process calls from the recordings. The resulting parameters are then used to identify the calls by means of statistical techniques. The actual implementation of the various systems may differ with regard to the parameters extracted, the statistical process used to identify the species, and general aspects such as user-friendliness and speed. These differences are not examined in detail here, as they only play a minor role in the context of a general description of automatic recognition.

6.1.1 The process of automatic identification

Automatic recognition starts with the search for calls within the recordings. There are various selection criteria, which include parameters such as wave shape and frequency. The latter are established by means of various processes such as zero-crossing analysis or Fast Fourier transforms (FFTs). Frequency characteristics as well as flanks are evaluated. The objective is to include as many of the 'good' genuine calls as possible in the analysis, while ignoring other signals, disturbances and calls of poor recording quality. In this way, the rate of failed identifications further down the line is reduced. The quality of the call search process has a strong bearing on whether calls are missed or whether large amounts of interference are carried on to the next stage, the identification process. Thus, this stage has a strong impact on the subsequent steps.

After the calls have been found, they are automatically processed, and parameters extracted which characterise and describe the call (Figure 6.1). Similar values to those from a manual evaluation are extracted quite independently of the system. To improve identification, some programs establish values which define the shape of the call or the call type numerically. Thus, anything between a few and more than 100 parameters are used for the identification. At least one program uses processes from image processing to analyse standardised call sonograms

The parameters obtained are subsequently fed into a statistical process in order to arrive at an identification. Experience has shown that a hierarchical structure is most appropriate for identification. Calls are not narrowed down to the species in one step, but undergo several tests, beginning with a determination of the family or genus. The algorithm continues working down the hierarchy towards the individual species, but it will not continue to each subsequent stage unless the data are good enough. Thus, the algorithm may stop prematurely at the family or genus level and never achieve a precise species identification.

In general, each individual call is investigated, and the calls are collated into one overall result for the whole recording. An identification based on several consecutive calls allows inter-pulse intervals and rhythm to be included in the process. However, it is not possible to be sure whether every call has been picked up, nor whether, in any given case, only a single individual has been recorded. These can only be determined with any certainty if the recording is manually processed.

Ideally, the result will be an identification that is 100% correct, but it has to be treated with caution because the statistical basis of the identification

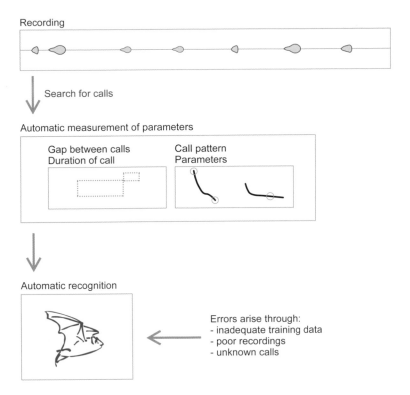

Recording

Search for calls

Automatic measurement of parameters

Gap between calls
Duration of call

Call pattern
Parameters

Automatic recognition

Errors arise through:
- inadequate training data
- poor recordings
- unknown calls

Figure 6.1 Automatic recognition of recordings involves the automatic extraction of parameters from bat calls. These parameters, which are used to identify species by means of statistical processes, may differ significantly between the different programs available.

does not draw on the full call spectrum of the species, but only on a training dataset stored by the software. Thus, the best that can be said is that the identification is probably correct, given these training calls (see also Figure 6.2 and Figure 6.3).

6.1.2 Differences between the systems

There are broad differences between the different systems which relate to the application and the underlying statistical analyses. The programmer has a very strong influence on the results, and the user needs to understand how they are obtained.

The user must be clear whether the software is based on a hierarchical or heterarchical (non-hierarchical) structure. In hierarchical systems, the complexity of the analysis and of the results is greater. The advantage is that the identification results are better if the training dataset has been adapted to the tree structure.

A further important aspect is whether there are differences in the weighting given to individual species or whether a simple variance measure defines the attribution to a species. There might, for example, be a bias against rare species, such as Bechstein's bat, which are difficult to identify but which have a great significance for nature conservation. Often the software will discount calls whose range overlaps with other species, in this case in the genus *Myotis*. This will result in the species not being identified.

In the case of the Bechstein's bat, for example, batIdent software will only give a confirmed species identification if there is fairly high degree of certainty. Otherwise the result is given as *Myotis small/medium-size*, because, as well as Bechstein's bat, this genus includes Daubenton's and whiskered bats, which have similar calls. This process is not carried out via the identification function, since it is already programmed in the training dataset through the selection of calls and by adaptations of the outlier analysis. A comparison of the systems using the same recordings reveals big differences in the error rate, and in the types of errors (Rydell *et al.* 2017).

There is also significant variation in the operating speeds of the different systems. This is an important consideration, since it is not desirable for the analysis to take several hours or days to produce results. Currently, the fastest system, batIdent, operates at a rate of about one second per recording, while other systems require considerably more time for the analysis. iBatsID evaluates and produces an identification result for each call in the recording. The user then has the time consuming task of working out manually the overall result for the recording. SonoChiro produces Excel lists of the results for the purposes of further processing. BatScope/BatExplorer and bcAdmin/batIdent, on the other hand, provide a user-friendly overview of the results and allow quick manual follow-up checking.

6.1.3 Limitations of the process

Reliability of identification is positively correlated with the number of high-quality calls (Pye 1993, Parsons *et al.* 2000). Poor-quality calls, as well as signals not originating from bats, lead more frequently to identification errors. For that reason, it is best for as many bad calls as possible to be picked out and excluded before the identification stage. This may mean that some recordings do not undergo the process of identification at all, and that some species may be missed. In practice, identifying bad calls involves a great deal of work and effort, since, for example, fragments of longer calls may be easily misinterpreted as short calls. Equally, echoes developing from the original calls pose quite a challenge. This problem can be circumvented by stopping the search for calls once a call of a certain length has been found. In this way, the measuring of echoes after the end of a call will be largely eliminated.

The quality of the training dataset of the predictor restricts the possible applications of the system. Just as the experience of a person using a bat detector is a decisive factor in manual identification, so it is with the training dataset used in the statistical process (Biscardi *et al.* 2004). Ideally, the training dataset will have included every call type used outside the roost by every species, and with a weighting which corresponds to the abundance of those bats in the habitat. Since this is impossible in reality, and assumes regional libraries of local bat calls, the possibilities of the system are limited by the quality and comprehensiveness of the training dataset. It is a common occurrence to record bat calls outside a roost which have not been included in the training dataset. By reducing the number of parameters to two, Figure 6.2 illustrates this in a very simplified way. Automated long-term recording has made it possible to investigate locations which it was once rarely, if ever, possible to survey. The large quantity of recordings will always include new or unknown call types. This will inevitably lead to errors in identification.

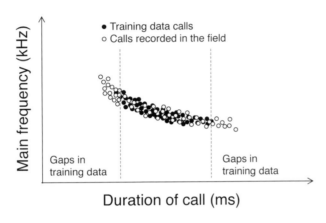

Duration of call (ms)

Figure 6.2   If, at a particular location, calls are recorded which are scarcely included in the training dataset or omitted entirely, the system will see them as unknown calls, which will lead to an increase in the error rate. It must also be borne in mind that manual identification is not always possible.

If an unknown call, that is to say, one not in the training dataset, is inputted, the system will either flag it up as *unknown*, or alternatively try to find another call in the training dataset which is as similar as possible. Flagging up a call as unknown reduces the failure rate significantly, but it will mean that numerous recordings made in the field will no longer be identifiable. For that reason, the latter solution, finding another call, is generally the one chosen, since this is easier to implement statistically. Nevertheless, it is sensible to use outlier analysis to examine calls which deviate from the training dataset, and to mark them in a suitable way. Alternatively, the predictor variable could learn unknown calls, but there is the risk of an initial identification error which could then compound errors further down the line.

As well as the uncertainties created by unknown calls, identification errors can arise from the overlapping of calls of different species, as illustrated in

Figure 6.3. If the calls of the bats present in the area are generally very similar, the system will make frequent errors. This is true not only for the species that overlap in the frequency range, but it is to be expected for nearly all calls. Depending on the distribution of the calls, error rates of 20–50% are quite possible (Figure 6.4).

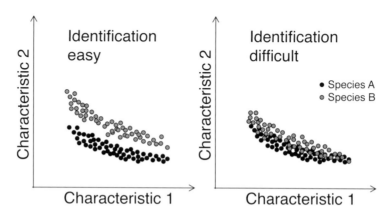

Figure 6.3   On the left, two species are shown whose calls do not overlap if two characteristics are considered. On the right, the calls show similar values and do overlap. This will result in incorrect identification.

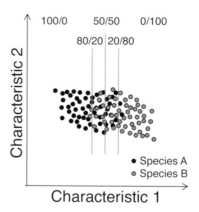

Figure 6.4   If the calls of two species are categorised according to pairs of parameters, the result will be some sectors without overlap and other sectors with increasing overlap. Error rates of 20–50% are then to be expected.

If there is an unfortunate choice of calls inputted into the training dataset, or if the recording takes place at a location where the calls are atypical, this may result in the misidentification of species (Figure 6.5). The problem is not only the identification errors themselves, but also the fact that it makes it difficult to assess the quality of an automated recognition system.

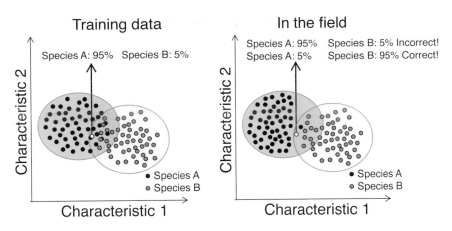

Figure 6.5   An unfortunate choice of reference calls or a location with unusual calls may greatly reduce the quality of the identification.

The provision of error rates, which are valuable to the user as a criterion for choosing a system, is therefore not entirely to be trusted. If one's own call results deviate from those of the test data, then it is reasonable to assume that the quality of the data may deviate in other circumstances too. It is therefore very difficult to give a single figure for the global effectiveness of a call recognition system. In order to make meaningful comparisons, a wide range of calls must be used for the tests.

Two other errors which may occur during evaluation must also be taken into consideration, namely false-positive and false-negative identifications. These errors are independent of each other and are best analysed using so-called confusion matrices. It is also possible to use these to determine the effectiveness and the sensitivity of a process. The effectiveness corresponds to the proportion of calls identified correctly as a particular species relative to the total number of calls identified (rightly or wrongly) as that same species. The sensitivity shows the share of calls correctly identified as a particular species as a proportion of the total number of the calls of that same species. Both figures can be used for assessment, but they cannot be transferred directly to recordings in the field. They are only valid for test calls, from which they were established. Effectiveness is particularly prone to large fluctuations. Even if species are absent from the test dataset, these may nevertheless appear in the results of the identification, and thus be 100% false.

Tables 6.1 to 6.3 are a simplified illustration of the effects on sensitivity and effectiveness of differing proportions of species whose call frequencies overlap. Table 6.1 shows the results with the same number of calls of each species and a normal overlap of 10%. Table 6.2 shows the result with a significantly lower number of calls of Species B without further change.

In Table 6.3, again, the same number of calls are shown, but with a higher overlap of Species A and B assumed. Figure 6.6 plots the calls in the three hypothetical tables 6.1, 6.2 and 6.3, showing two parameters in a simple feature space.

Table 6.1 Distribution of the calls of two species in a hypothetical feature space, with the resulting confusion rates. There is the same number of calls for each species and a good representation of call variability.

Prediction True	Species A	Species B	False negative Sensitivity	N
Species A	450	50	10%	500
Species B	50	450	10%	500
False positive Effectiveness	10%	10%		

Table 6.2 As with Table 6.1, but with significantly fewer calls of Species B. The effectiveness changes.

Prediction True	Species A	Species B	False negative Sensitivity	N
Species A	450	50	10%	500
Species B	20	180	10%	200
False positive Effectiveness	4.26%	21.74%		

Table 6.3 As with Table 6.1, Species B has many calls in the overlap area. There is a clear change in effectiveness and sensitivity.

Prediction True	Species A	Species B	False negative Sensitivity	N
Species A	450	50	10%	500
Species B	150	350	30%	500
False positive Effectiveness	25%	12.50%		

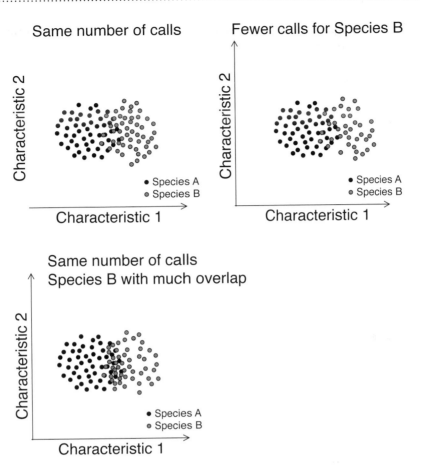

Figure 6.6   Calls in a simple feature space with two parameters. The diagram clarifies the datasets in the confusion matrices depicted above.

6.1.4 Advantages of automatic call analysis

Automatic call analysis has proven to be an indispensable technique in many current studies. Only with this technology is it possible to cope with the large number of recordings from passive monitoring. A mere few weeks is enough to produce up to 10,000 recordings. The time consuming and expensive manual processing of all the recordings is rendered unnecessary with automatic call recognition, and the results obtained are not compromised by subjective errors. The recordings can also be checked manually, for example in order to eliminate obvious errors.

As illustrated in Figure 6.3, individual species can overlap so significantly in their calls that errors will inevitably arise. The statistical identification of species allows a better separation, as up to 100 parameters can be included, so reducing the overlap between species. Human beings, by comparison, are not able to process more than a few parameters per call. On the other hand, not all identification errors will necessarily be eliminated, as many of the measurable parameters are not independent of each other. The bandwidth of a call, for example, is dependent on the start and end frequencies.

Since the statistical process always makes the same objective errors, these can be followed up subsequently, even by third parties. Thus, after the analysis, it is possible to evaluate the rate of false identifications, and which recordings will contain these errors. Because of the objective procedures of the automatic predictors, it is now possible to achieve a genuine data comparability from various recordings.

With up to 10,000 recordings from long-term monitoring, it is an enormous advantage if the software is fast, ideally as close as possible to one call per second. When initial data results are available promptly after the recording, it becomes possible to scan through them for interesting findings.

6.1.5 Disadvantages of automatic call analysis

As discussed above, the training dataset has a significant influence on the results. Furthermore, the automatic system can completely fail in some situations. For example, if at a particular location there are many recordings of calls which are scarcely represented in the training dataset, or completely missing from it, there may be high error rates. Statistical species recognition generally produces fuzzy results, and it has to be assumed that there will always be a certain imprecision because of false-positive and false-negative results. Error rates are always species- or situation-specific, so that a blanket interpretation of the errors is not always feasible. Because calls are analysed individually the chronological sequence of calls within a recording is ignored, which means that some components that may be critical for identification are not considered. Likewise, it will not be possible to integrate the data from several recordings, a technique which can help to improve the identification results. However, by switching to manual control after the automatic process, the chronological context can be used to make rapid corrections.

6.2 Critiques of automatic systems

Many authors consider that incorrect species recognition by these systems will lead to wrong decisions being made regarding species management or impact studies. They demand a significantly more comprehensive, independent validation of the systems before they are deployed more widely. However, automated systems have been in use for many years and have produced large quantities of data. It is only through these systems that the data needed for species management can be obtained. Without automated analysis, these data could not be used meaningfully, and there would be negative consequences for species protection. Users are able to do a follow-up check on the results, and it is very much their responsibility to carry this out with species that are rare or difficult to identify. It must equally be stressed that some species are not easy to identify using manual techniques either, and that the user has a significant effect on the results.

Firm guidelines for acoustic recording, drawn up by independent experts, are long overdue. Users of acoustic methods could then undergo training based on these guidelines and be issued with a certificate in recognition of their training. Similarly, a scheme for the use and evaluation of automatic recognition software could be established. Within the framework of such guidelines, the limitations of acoustic studies could be listed, for example relating to species that have quiet calls or are difficult to identify. These could then be drawn upon, irrespective of whether the methods are automatic or manual. In this way a standard, uniform analysis would be possible.

6.2.1 Correction of the results of automatic recognition

Users need, moreover, to respond to the critical voices aimed at automated species recognition systems, for example by systematically checking and correcting cases of misidentification. This is possible with moderate effort and contributes to a significant improvement in the end result. It is recommended that a methodical process be adopted, based on the errors that are regularly produced by identification software. Users can justifiably be accused of negligence if they extract data directly from automated systems without undertaking proper checks.

Automated recognition systems based on a statistical process will always commit the same objective errors. However, that means that it is possible to carry out an effective check and a manual correction of the results if these errors are known. These types of errors are linked to the range of species and the precise location. If the same system is always used, it is easy to become familiar with the typical errors, correct them and improve the quality of the results. Using batIdent in combination with bcAdmin, a system regularly operated by the authors, an explanation is given of possible procedures for improving the results of large quantities of data of up to 10,000 recordings. An indication is also given of the time requirements and its impact on outcomes.

6.2.2 Effective correction of the results of large quantities of data

For the following steps to be carried out most effectively, it is recommended that the recordings, as well as the species dataset entries, are managed in a databank or on a similar software platform. Ideally the recordings can also be sorted or filtered according to criteria such as date and duration of recording and species. A short description of the calls in the program is helpful and speeds up manual checking. There is a large loss of efficiency in the checking if there has to be a lot of flipping between folders of recordings, analysis programs and tables. Figure 6.7 portrays the process schematically.

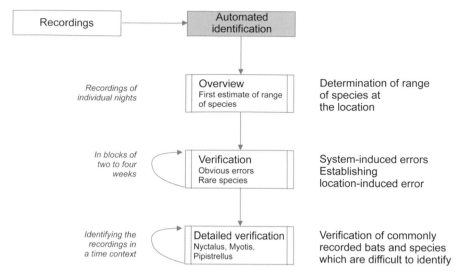

Figure 6.7 Schematic overview of the procedures for enhancing the results of automated identification. This is designed for the processing of large quantities of data, for which manual verification of the individual recordings is too time consuming.

After the automatic identification of the recordings has been carried out, the first overview of the range of species will become available. It is now a good idea to do a spot check by selecting individual days and verifying the results manually. This will allow a first impression of the species encountered and any potential misidentifications. These may be specific to the location or region and transferable to other locations in the same area.

Using the first impressions as a basis, the next step is to correct the obvious cases of misidentification in time blocks of 2–4 weeks. In this way, the typical error and misidentification patterns for the area will be progressively narrowed down. The aim of the first manual correction is to discount species that do not occur in the region at all. At the same time, there should be a check for rare species or those which are seldom recorded. In general, a large number of misidentifications will already have been eliminated once these first steps are completed.

Next, a check is made for commonly occurring species that are difficult to identify. Above all, this means *Nyctalus* and *Myotis* species, especially the latter if they were not already checked for in the first step. The species that have to be checked are based not only on those that occur at the site, but also on those that are subject to frequent misidentification by the software employed. For effective verification it is essential to consider the recordings in their time context. Recordings which were made within a short time span should be considered together for this purpose. The assumption will be that they originate from only one individual. If it has been possible to reliably identify a recording in the same time context, this species identification can be applied to other recordings. In this way, large quantities of recordings can be processed very quickly, even if using this method does involve a few misidentifications. A value of between 10 and 30 seconds is suitable for the sequence of recordings. If there is a significant occurrence of high-flying species, a longer time window should be chosen.

By using this procedure, it is possible to achieve extremely consistent results, even if a few errors are to be expected. Indeed, even laboriously checking all the recordings manually would not necessarily achieve a better result.

The typical procedure employed is illustrated below by the example of a location in the city of Münster, where bat activity was continuously recorded from May to October. It was expected that there would be significant activity of Leisler's bats as there was a nearby maternity roost. Other species regularly recorded in urban areas are noctule, serotine and common pipistrelle. It was not known whether any other species were present. These species represent the typical range for a town area, whereas woodlands would generally include a stronger contingent of *Myotis* species.

In general, it is recommended that data are corrected in blocks of between two and four weeks in duration, as the range of species is subject to few changes over such a period of time. If larger blocks of data are checked in one go, the verification will be more subject to errors due to factors such as seasonal change. Shorter, more closely associated data streams can then be checked according to a simple formula. In the following example, two recording blocks are presented, from 01/08 to 15/08 (3,246 recordings in Block 1) and from 15/09 to 30/09 (3,023 recordings in Block 2).

Initially, checks were made of the recordings of species that were only very rarely encountered. For Blocks 1 and 2, the bats verified were: barbastelle (shown as Bbar in Figure 6.8), northern bat (Enil), all *Myotis*, Kuhl's pipistrelle (Pkuh), low- and medium-frequency pipistrelles (Plow and Pmed respectively), pipistrelle species, Schreiber's bent-winged bat (Misch), Savi's pipistrelle (Hsav), and parti-coloured bat (Vmur). For the purposes of rapid processing, the list of all the recordings in the time block was filtered, so that only the relevant recordings were shown. This summary was then scanned through, and the cases which required precise analysis were uploaded into an analysis program. In this way, the first misidentifications were spotted and corrected. The processing of the two blocks took about 15 minutes each, and about 15% of the records were corrected. The number of species was thus reduced from just over 20 down to between 12 and 14.

In a second step, the nyctaloid species (nyctaloid, Nlei, Nnoc, Eser) and the pipistrelloid species were checked. The second verification is normally more time consuming, even if, by using the time context of the recordings, not all of them have to be verified individually. As discussed above, not all recordings which follow each other in quick succession (for example at intervals of 10 or 30 seconds) are verified but are attributed to a single individual. This is achieved by the verification of a few easily identifiable recordings. This second step was completed within 20 to 30 minutes for the two blocks investigated. Figure 6.8 shows the change in the attributions of the species as a result of the first, second and concluding verifications.

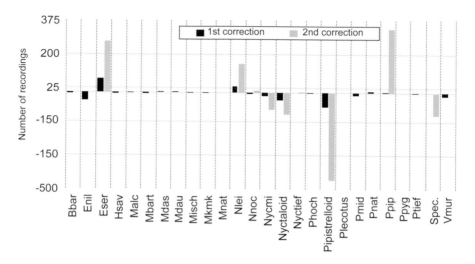

Figure 6.8 Illustration of the improvements in identification accuracy that can be achieved by verification and correction. The graph shows the changes in the numbers of each species identified as a result of the first and second corrections.

The optimal procedure depends, to a certain extent, on where and how the errors come about. It therefore varies according to the system used. However, with a bit of practice, analogous procedures can also be found for other software packages. In the example above, 30% of the recordings were able to be corrected within 40 minutes. The proportion of errors remaining due to incorrect identification is extremely small. There is a high degree of certainty that data for rare or non-native species will have been corrected. Manual identification of all the recordings would have taken up to 100 hours.

6.2.3 Procedures for recordings with significant interference

It is by no means the case that long-term recording will only monitor bat calls. A high density of bush-crickets, for example, or a large amount of interference from the nacelles of a wind turbine will result in a greater number of false-positive recordings. The automatic recognition system may confuse these with bat calls when the call signature looks similar. Checking will require a great deal of effort if there are tens of thousands of recordings.

One possible solution in this case is the sample testing of, say, one in ten recordings. If bat calls are found, the adjacent recordings may also be checked. This procedure can also be supported by suitable software. This is really the simplest solution when dealing with large quantities of data. Care must be taken that the time gaps between recordings are not too long, as there is then the danger that bats will be missed.

Alternatively, it may be that interference and bat calls can be distinguished by simple parameters. The recording times should always be used as a first criterion. Bats are not generally active in the day, so any such recordings can be eliminated. Before deleting any recordings made in half-light, up to an hour before sunset or an hour after sunrise, they should be checked in case they do contain bat calls.

Sometimes the length of the recording itself is a helpful distinguishing feature. After sorting the files according to duration, both short and long recordings should be checked for bat calls. Ideally, these will be found in only one of the two categories. Interference will mostly be found in the short recordings, while call sequences with several calls mostly produce significantly longer recordings. Nevertheless, such a separation is not always so straightforward in practice.

Another possibility is to use the results of the call search. Recordings with no calls suggest the absence of bats, but even those which do contain calls can be sorted by the identification result and verified. So, for example, there may be interference that, because of its frequency range, is identified as horseshoe or as the social calls of pipistrelles. In order to determine such patterns, it is a good idea initially just to check single nights. If a distinguishing feature can be found, this can be applied to all the data.

With a suitable procedure applied to all the available recordings, even very large quantities of data can be analysed within a reasonable time and cost-effectively. Under certain circumstances, compromises over the quality of the data will have to be made, as there may be an error rate of up to 10–15% with the procedures described here.

7 A comparison of identification methods

Which is the best method of identification? It is a question often raised, but is there really a *best* method? Or are manual and automatic recognition techniques equally valid, since errors are primarily determined by the variability of the echolocation calls? It also raises the question of which parameters should be taken into consideration for the purposes of comparison. How can *better* decisions be made?

Table 7.1 shows the identification methods introduced in previous chapters in simple form, and compares them with reference to the characteristics discussed. It will be helpful here to recap briefly on the meanings of some of the important terms:

Identification in the field	always refers to the use of heterodyne or frequency division detectors.
Manual identification	the evaluation of recordings with the help of sonograms.
Automatic identification	an umbrella term for all the autonomous systems for identifying species.
Influence of the user	the impact of experience on the quality of identification.
Context	covers the role of time and behaviour in aiding identification.
Reproducibility	the extent to which consistent identification results are obtained when a procedure is repeated.
Checking	Quality control of the results of an identification process.

Identification by ear, using bat detectors in the field, will not be discussed any further here. It constitutes a special case because the accuracy of identification in the field can only be assessed subjectively.

With manual identification, the level of experience of the user has a strong impact on the results. Experienced users are mostly better than automatic systems. However, the human mind is limited in the amount of data it can process, and considerable time is needed for an informed analysis.

Even if automatic systems can identify species in a short time and with a reasonable degree of accuracy, they can never completely replace the human operator. As we pointed out in Section 6.2, *Critiques of automatic systems*, the combination of the two methods is the most effective way.

Table 7.1   Comparison of the characteristics of various identification methods

	Identification in the field	Manual identification	Automatic identification
Influence of the user	high	moderate	low
Context	yes	varies	varies
Reproducibility	low	moderate	high
Checking	no	yes	yes
Comparability	low	low	high
Speed	carried out directly in the field	low	high
Large data quantities	no	no	yes
Probability of correct identification	not possible to quantify*	not possible to quantify*	varies*
Costs	high – staff	moderate – staff	low – software

*A computer is able to give an objective figure for the probability of a correct identification, whereas a human being can only work subjectively and cannot give a reliable figure for the probability.

7.1 Is better identification possible?

When discussing the pros and cons of identification methods, it must never be forgotten that it is the bats themselves that make the identification of their echolocation calls easy or difficult. Identification can only be as good as the underlying reference data. If two species use calls which are similar or even identical according to the current technical capabilities of our recording technology, then distinguishing between them is no longer possible (Figure 7.1). Only when further parameters, not extracted directly from calls, are drawn upon, is it perhaps possible to make meaningful judgements.

Parti-coloured   Serotine bat
bat

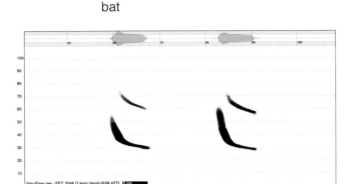

Figure 7.1 This sonogram shows an example of overlapping calls – those of the parti-coloured bat (left) and the serotine bat (right).

The human mind can theoretically achieve better results in this situation than an automatic system. However, a human operator cannot guarantee or calculate the probability of a correct identification. Furthermore, the human mind makes mistakes and cannot identify every call sequence of every species. Bats will also very occasionally make calls which are unknown to us. These are often difficult or impossible to identify.

The identification of calls becomes progressively more difficult for both human and machine as the recording quality worsens. Poor-quality calls may not be identifiable by either manual or automatic means, irrespective of whether the issue is the characteristics of the device, the low volume of the recordings, calls drowned out by interference in the surroundings, or the echoes from bat calls. In Chapter 8, *The complexities of call analysis*, the potential causes of poor recording quality are presented. We also discuss problems which can result from the transmitter, sound propagation, digitisation and the measuring of calls. If these aspects are taken into consideration during the design and construction of devices, the accuracy of identification will improve due to better recording quality.

Even the location has an impact on the accuracy of identification, both for automatic and for manual methods. If locations are chosen where the species tend to use typical calls that are distinguishable from other species, then the accuracy of identification will increase. Conversely, recordings from roost sites which are enclosed flight spaces tend to contain many echoes as well as numerous short calls that are difficult to recognise. Accurate identification often becomes impossible. In most cases, the location cannot be freely chosen, as it may be integral to the study. On the whole, however, a certain flexibility will be possible, and it should be used to improve the precision of the identification.

In practice, it will, of course, never be possible to identify every call in every recording. Low recording quality and overlap of calls between species are the principal reasons for this. Through optimisation of the technology and construction of the equipment, and by targeted selection of locations, the results can often be improved. However, there will always be a few recordings which cannot be reliably identified. Irrespective of whether manual or automatic methods are used, a meaningful approach to these recordings must be adopted. If it is definitely possible to attribute a recording to a genus or a particular category of calls, then that is a better result than an incorrectly identified species record, even though the precision is lower.

Nowadays, a huge number of acoustic recordings are being made. Before 2010, little of this kind of work was carried out, but since then the number of users and systems has climbed steeply, above all because of the availability of modern technology. It is likely that around a terabyte of data is being collected every day for this purpose. Rarer species are now being more frequently recorded than a few years ago. Knowledge of echolocation is still being extended, and this is leading to more accurate identification in the medium term. All this is helped by factors such as a higher proportion of identifiable recordings and a better awareness both of potential errors and of the calls which cannot be identified.

8 The complexities of call analysis

The identification of bat calls from sound recordings is not always easy. Quite apart from the overlap of calls between species – in other words, the biological similarity between calls – we must also take into account the technical characteristics of the analytical tools and the recording technology. The transmitter, namely the bat itself, as well as the propagation of its signal, also have an impact on the recorded call. Chapter 5, *Manual identification of species*, dealt briefly with some of these problems. The current chapter will delve more deeply into the principles of manual identification on the computer. There will also be a discussion of the various factors that can distort recorded calls.

8.1 Characteristics of the transmitter

When a bat transmits a call, it decides on the structure. Bats always seek out an optimal solution to the task at hand, whether it is navigation or seeking out prey. To achieve that goal, the bat adapts its calls to the prevailing conditions. Age and gender may also have an influence on the calls, although that sort of information is not usually available to the user of the detector equipment.

If the bat is moving relative to the microphone while it is calling, then there will be a Doppler effect, an everyday phenomenon we are familiar with from hearing ambulance sirens. A sound from a source that is moving towards the observer seems to change to a higher frequency whereas, when it is moving away, it seems to change to a lower frequency. The transmitter maintains a constant frequency throughout, and it is the movement that brings about the change in the sound waves. Bat echolocation calls behave in the same way. The recorded frequency will shift according to the relative movement. A bat flying at 8 m/s, for example, will cause a Doppler shift of around 2%. A call of 45 kHz emitted by a bat flying towards the microphone will shift in frequency to 46 kHz as the bat passes. For the same reason, fast-flying bats may cause errors of up to 2 kHz.

As well as the frequency range and duration of the calls, the sound lobe can also be altered by the bat. The beam of sound can be widened or narrowed and more precisely directed. If a narrow-beam call is not directed at the detector microphone, the recording will be faint or non-existent. This behaviour will therefore have a direct impact on the distance at which the call can be picked up by the microphone. When the bat narrows the sound lobe of the call, then it is much more likely to be missed (Figure 8.1).

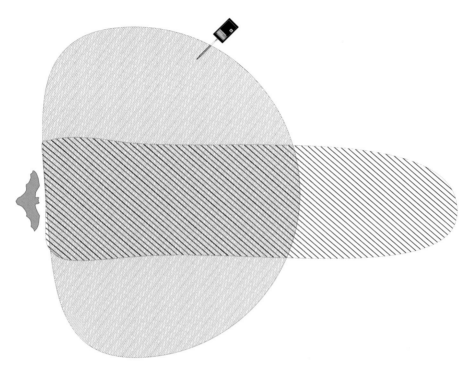

Figure 8.1   The bat varies the sound lobe of its call and therefore affects the likelihood of the call being picked up by the microphone.

The sound lobe of a bat is generally not symmetrical, but tends to resemble a searchlight beam. It will vary according to the particular activity, being modified, for example, for a close-range feeding buzz. Although there has been relatively little field research on the typical shape of the sound lobe, evidence seems to suggest quite a narrow opening angle and a rapidly descending intensity upwards and to the side (Figure 8.2).

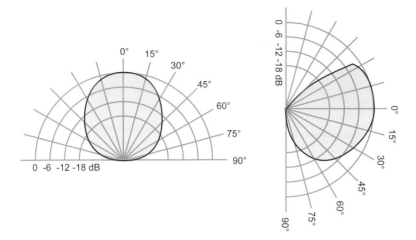

Figure 8.2 On the left is the horizontal spread of the sound lobe (seen from above) and on the right is the vertical spread (adapted from Jakobsen *et al.* 2015).

With broadband FM calls, the span of the sound lobe acts like a frequency filter. The acoustic spectrum of the call directly in front of the bat differs from that at the margins of the sound lobe. The bat can directly modify the frequency characteristics, by amplifying or attenuating certain frequency ranges. Particularly in the case of species with frequency-modulated calls, this can lead to the recorded calls varying significantly in their spectral composition and in their main frequency.

8.2 Factors impacting on sound propagation

The call, which travels the distance from the bat to the microphone as a sound wave, is subject to a variety of physical factors modifying the call. Well-understood phenomena such as absorption or attenuation cannot in practice be modified nor, in the case of wind and layers of air, can it be measured in any meaningful way. On the whole, then, no correction, and thus no general evaluation of the effect of such factors, is effectively possible.

8.2.1 Attenuation through atmospheric factors

Atmospheric conditions have a strong impact on modulated calls, and increasingly so as the frequency rises. This means that the propagation of higher-frequency calls is hindered more than that of lower-frequency calls. With frequency-modulated calls, the microphone will only pick up the lower-frequency parts of the calls if the bat is further away (Figure 8.3). This is discussed in greater depth in Section 14.1.3, *Propagation of sound*.

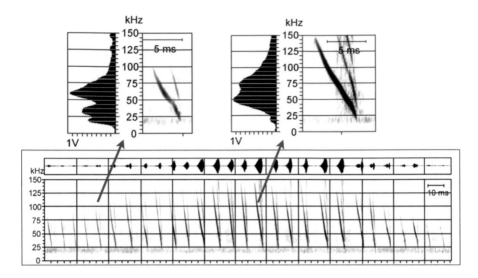

Figure 8.3   Because of the attenuation of the call through atmospheric factors, higher frequencies propagate less well than lower frequencies. With FM calls, this can mean that parts of the call are not detected as the bat moves away.

8.2.2 Multidirectional spread and reflections

Another effect that impacts on the recording is the multidirectional spread of sound waves. Because of reflections off objects, one or more echoes may arrive back at the microphone as well as the actual calls. These will arrive with a time delay if the paths taken by the sounds differ (Figure 8.4). For the purposes of estimation, the speed of sound is 340 m/s, or three milliseconds per meter. This allows a crude estimation of the risk of echoes when a microphone is being built.

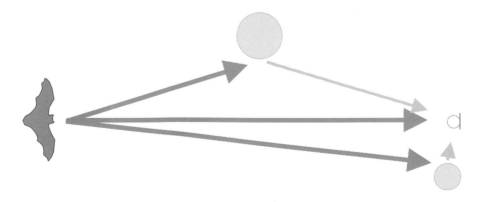

Figure 8.4   Multidirectional propagation of sound causes an overlap of calls and echoes.

If the time lag of an echo is shorter than the duration of the call, overlapping will occur. In the worst case, interference will result from this. The so-called beat frequency can cause parts of the call to be cancelled out. When the sound waves are out of phase (antiphase), the call will be cancelled out, whereas when the waves are in phase, the call will be amplified. Spectral changes will also occur, and will make identification of species significantly harder. Even the measurement of the length and direction of the wave is not always possible. Echoes can never be completely avoided, but the likelihood of strong echoes occurring can often be reduced by choosing a suitable location. That, however, mostly means positioning the microphone at a distance from the recording device (see below).

An example of this is recordings over water of foraging Daubenton's bats. Parts of the signal are eliminated as a result of interference produced by echoes arriving almost at the same time as the original calls. This phenomenon was for a long time a good identification aid for this species (Figure 8.5). Water acts like an acoustic mirror and reflects the calls of low-flying bats almost entirely away from them. Because of the bat's low height above the water and the resulting short time gap, the call and the echo arrive almost simultaneously at the microphone. This leads to strong interference.

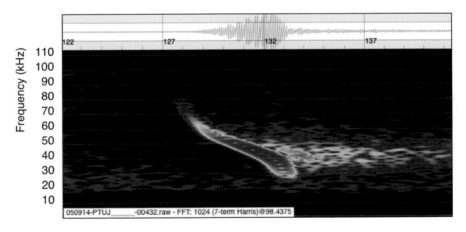

Figure 8.5 The call and echo of the Daubenton's bat are superimposed when it is hunting low over water. The effect of the beat frequency can be recognised as gaps in the call pattern.

Reflections of the sound can also occur in other species and situations in the field. Longer bat calls, for instance, will make it more likely that the echo will be superimposed on part of the call. This can only be avoided if both the microphone and the bat are located far enough from the reflecting objects.

An example of this is seen in the case of the noctule, which is almost always recorded with slightly superimposed calls, since the detector is either in the hand or on a short, 2–3 m mast (Figure 8.6). Hard sound-reflecting horizontal surfaces, such as water, roads or firm ground, will cause the microphone to pick up strong echoes. Even vertical surfaces such as bushes will reflect calls and lead to superimposed signals.

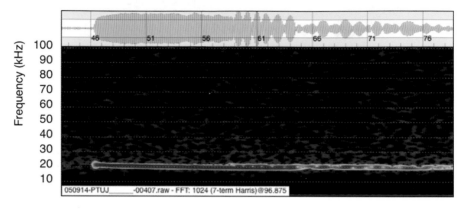

Figure 8.6   Overlapping call and echo of the noctule. The end of the call is no longer clearly recognisable. The interference patterns are similar to those of the Daubenton's bat over water (see Figure 8.5).

Overlapping signals of this type will make it difficult to recognise both the end and the duration of the call, and spectral measurements become problematic. As the strength of the echo and the degree of overlap with the call increase, the extraction of identification parameters becomes progressively more difficult. Since sound travels at about 34 cm/ms, these types of reflective surfaces can have an effect on the recordings of longer calls, even if they are only 3–5 m away.

With careful planning and preparation of the equipment, it is possible to reduce the effects of strong interference to a minimum. By placing the microphone at a sufficient distance from the reflecting structure, the recording quality can often be markedly improved. Even a 3-m distance from the wall, ground or foliage barrier will mean a time lag of over 10 ms, which is usually quite sufficient to separate the call and its echo. With many detectors, however, the fact that the microphone is built in or attached will mean that strong interference and reflection of sound could pose a problem. Echoes develop on the surface of the device, and having the microphone positioned close to this surface will lead to massive overlapping (Figure 8.7). In this situation, an optimal setup in relation to other structures will not alleviate the problem at all.

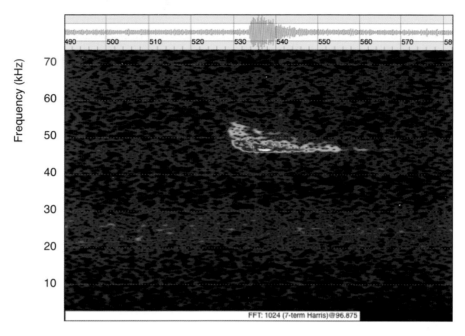

Figure 8.7 An inappropriately fitted microphone can result in recordings which are so badly superimposed with echoes that identification becomes almost impossible. The image shows a call which is probably from a common pipistrelle.

8.3 The impact of the recording technology

The sound recorded by the microphone is first amplified and filtered by analogue circuitry. This stage depends to a large extent on the quality of the processing of the sound waves. For the storage and further processing of the data, the analogue microphone signal is then digitised.

The previously analogue signal is reproduced with a constant sampling rate in discrete voltage values, in order to store the sound waves together with the volume. This means a significant decrease in data, which can result in changes in the signal (see also Section 14.6, *Digitisation*). The digitisation affects not only the frequency range, but also the likelihood of retaining faint signals (because noise has been eliminated). If amplitude resolution and time resolution are low, the analysis of the signal is rendered more difficult, and separating and identifying some species is then no longer so easy. It is particularly higher-order non-linearities in the circuitry which will lead to distortions in the input signal.

High noise levels, and thus poor signal-to-noise ratio of the detector, certainly make call analysis much harder (Figure 8.8). High noise levels can be produced by poor circuits with high internal noise, and by high amplification of the signal. A high amplification of the microphone signal amplifies not only the input signals, but also the noise inherent in the system. Quiet calls are thus hard to pick out, either in spite of or because of the high amplification

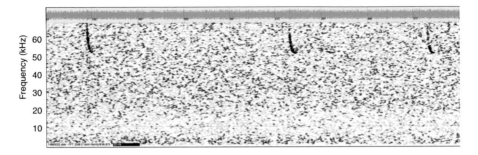

Figure 8.8   If the recordings have a bad signal-to-noise ratio, calls are difficult to analyse, as they are barely distinguishable from the noise.

8.4 The impact of the analysis

Ultimately, the call is recorded in order to undergo an analysis on the computer. To achieve an identification, call parameters are extracted from the recordings and compared with reference works as well as with the user's own reference recordings. Graphical representations of the wave shape and the spectrum are computed. The parameterisation of the process employed (FFT) has an effect on the data obtained. As is described in Section 14.7, *Representation of sound*, the size of the FFT, the type of window and the overlap all affect the results obtained. Frequency measurements with different settings can result in a deviation of one or more kilohertz (Figure 8.9). Even the software used can have an influence, since it determines the exact implementation of the computations. Whether spectra are established in one go, or are worked out by fragmenting (windows) and averaging, different results will be produced for the main frequency, above all with FM calls. For this reason, the settings and software used should always be documented.

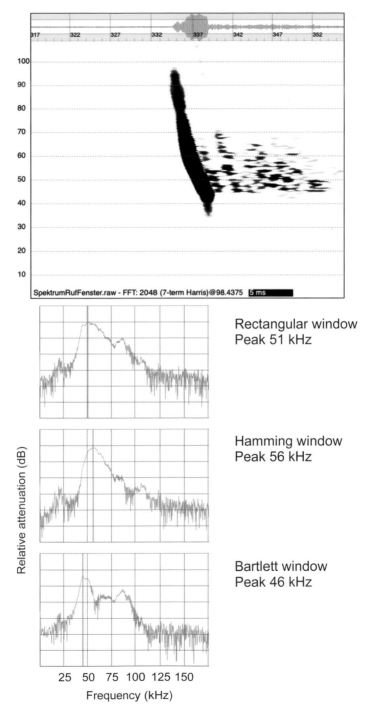

Figure 8.9   With otherwise identical settings, the main frequency of a *Myotis* species call will vary by approximately ±5 kHz according to the FFT window type chosen (Rectangle, Hamming, Bartlett)

8.4.1 Choice of parameters

In order to identify species, characteristic parameters of a call are extracted and compared with values gained empirically or found in existing literature (see also Section 5.2.1, *The process of identifying species*). Some of these key values are clearly defined and easy to pick out, whereas others are used without precise definition, and so are subject to very different interpretations by different users. If there is doubt, therefore, because of the lack of a clear definition of a particular parameter, the precise method of measurement and even the settings and type of software used should be documented. Otherwise, any data obtained from a study may not be comparable with other published results. The most important parameters are introduced and defined below.

Start frequency

This is the frequency at the beginning of the call. With modulated calls, it is often difficult to determine, as high frequencies do not carry as well because of the attenuation as the sound wave spreads. The angle of the bat to the microphone also has a significant impact on the frequencies recorded. As a result, the start frequency will not always be reliably recorded and may be missing altogether from the recording. Therefore, this value should always be treated with caution. It is, nevertheless, a useful feature for some species of *Myotis*, if it can be picked out in the recording. For other genera, the start frequency represents a less significant feature. It is true of all species, except the horseshoes, that shorter, modulated calls mostly start with a higher frequency.

Maximum frequency

The highest frequency during a call. Generally, this coincides with the start frequency. As with the start frequency, this value depends to a great extent on the quality of the recording.

End frequency

The frequency at the end of the call. This is often difficult to measure with some FM calls and when echoes are overlapping with calls. It is an important identification feature for numerous species with QCF calls. It is generally easy to measure, but with longer calls, and calls with quiet endings (often the case with *Myotis*), echoes can interfere with the measurement.

Lowest frequency

The lowest frequency of a call. Generally, this coincides with the end frequency. It is an important identification feature for numerous species with QCF calls. As in the case of the end frequency, it is a useful measure.

Bandwidth

This is defined by the frequency range of the call from the lowest to the highest level. The value is reliant on the successful recording of the high-frequency components of a call, and is only valid if the call has been recorded in its entirety. It is mainly used as a feature for identifying bats with FM calls.

Main frequency

Also known as *mean frequency* or *peak frequency*, it is measured as part of the spectrum. It is the frequency of maximum energy, and is useful for the identification of FM-QCF calls. With FM calls, the measurement is dependent on the frequency response of the microphone as well as on the directional accentuation by the calling bat or by the greater attenuation at higher frequencies. For these reasons, the main frequency should be treated with caution. In batIdent this measurement corresponds to Fmod.

Frequency at knee

The frequency at the point in the call where the first steep FM segment changes into a flatter FM or QCF segment (turning point). Sometimes this frequency is helpful for identifying *Myotis* bats.

Myotis kink

The frequency of the kink or bend (turning point) at the end of *Myotis* calls, where the somewhat flatter middle segment changes into the end phase or tail of the call.

Characteristic frequency

The frequency at the flattest part of the call. For QCF calls, it is useful for identification. In general, it the same as the value of the main frequency.

Duration of call

This is the length of the call, and is usually measured in units of time on an oscillogram. Because of the time resolution, which is determined by FFT, it is not possible to establish duration precisely using a sonogram. When the call has echoes superimposed on it, it is very difficult to measure. In the case of FM calls, the measurement is often too short, as the higher-frequency parts of the call are missing. It is usually unrelated to the call location, and it tends to be adapted to the echo density of the surroundings. Therefore, by itself, it is not a good identification feature.

Inter-pulse interval

This is the time interval between two consecutive calls, measured from call start to call start, and not to be confused with the call gap. The inter-pulse interval is used as an identification feature by some authors. However, since bats vary their calls and rate of calling according to the environment and to the purpose of the call, this measurement is not always useful. Nevertheless, it does help in the identification of serotines if Leisler's bat is also present in the area.

Call gap

This is the time difference between two consecutive calls, and is measured from the end of the first call to the beginning of the next call. As a rule, it barely differs from the inter-pulse interval, which includes the call duration.

Figure 8.10 summarises the above-mentioned parameters in graphical form. Most measurements are taken from the sonogram, although the main frequency can only be taken from the spectrum. Therefore, there are possible sources of error which may result from varying FFT settings. Usually, only call duration, inter-pulse interval and call gap are measured in the wave form graph. When extracting parameters, the graphs must be chosen so that the measurements are feasible. That means that the compression and zoom must be chosen appropriately, in order to be able to ascertain the parameters. It must be realised that the measurability of the parameters will be affected by the methods used in creating the spectrograms and sonograms (see Section 14.7.2, *FFT and spectrum*).

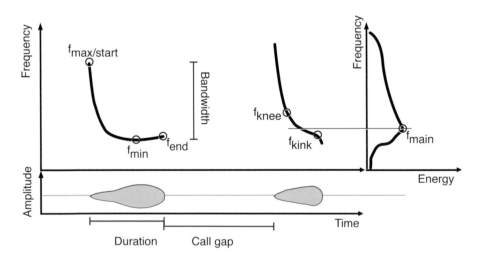

Figure 8.10   Graphical representation of the parameters discussed in the text.

8.4.2 Comparison with values in the literature

Since the spectral measurements are particularly influenced by the software and the settings used, comparisons with the literature need to be undertaken with caution. If authors do not specify precisely how the parameters were obtained, then calculations can lead to inconsistencies and lack of comparability between sets of data. The effects of calculating the spectrum are then particularly significant in the case of FM calls, but much less so with QCF or CF calls. Ideally, projects should specify the settings and the software used to obtain the values. As a rule, there is a lack of rigorous specifications in reference works. When a comparison is made of various standard works, it is obvious that the frequency parameters also fluctuate considerably. This is due less to the effect of the software used than to the recordings available. Both the choice of the call situations and the technology employed are reflected in the values obtained.

By way of example, Figure 8.11 shows two parameters, start and end frequencies, for the common pipistrelle taken from various widely used reference works (left-hand graph). The wide scatter in the start frequency is immediately apparent. This is presumably due to a number of factors such as choice of call, and the technology used. It is generally the case that higher frequencies will show a wider spread of values due to variations in measurement. The end frequencies of the calls, which are usually drawn on for identification, show a significantly lower spread of values in the comparison of data from the literature. Even here, however, there are clear differences. This becomes evident when one takes the final and main frequency values of the three species of pipistrelle (common, Nathusius' and soprano) and places them next to each other (right-hand graph). It is obvious that identification will vary according to which reference work is used.

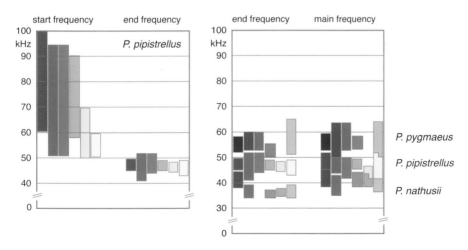

Figure 8.11   A comparison of the identification literature as exemplified by the start and end frequencies of the common pipistrelle (left-hand graph). Each bar corresponds to the values obtained in a reference work. The start frequency fluctuates considerably, although the end frequency shows fewer differences. When a comparison is made of the end and main frequencies of various species of pipistrelle, significant discrepancies can be seen in the literature (right hand graph).

It clearly follows that caution is needed when using reference works. Typical calls which are not in the overlap zone between species can usually be reliably identified. Where there is an overlap in values, however, this is not the case. The accuracy of identification can often be improved if a judgement is made on the basis of several parameters rather than just one. However, even three parameters can be quite a challenge, as truly multidimensional models only succeed with the help of special statistical processes, such as are used for automatic species identification.

9 Criteria for detector systems

For anyone starting out with acoustic and, particularly, automatic recording, it immediately becomes apparent that there is a very wide choice of devices and systems. The ideal solution would be to use a combination of devices, the best one available for each separate task, namely detection, recording and identification. However, it is rarely as simple as that. If, for instance, the identification software does not coordinate well with the recording software, the result can be far from ideal. By comparing the individual solutions for detection, recording and identification, the aim of this chapter is to shed light on the compatibility of the different solutions or systems. This is not straightforward, as there are a multitude of possible combinations. In general, however, it is true to say that:

- Even the best software is useless if the recordings include too much interference.
- Even the best detector will not perform if the required calls are not included in the software's training dataset.

There is now such a huge variety of solutions that it is difficult to keep track of the field. Furthermore, there are no clear definitions of the criteria for choosing a new system. Particularly when planning a new project, it is hard to predict which aspects are going to be significant. Anecdotal evidence is too vague, and test reports may often be irrelevant to the actual work envisaged. The important thing, therefore, is to consider carefully what goals need to be achieved and what tasks the system must accomplish. Decisions must be guided by the requirements and the modus operandi of the project, and, above all, by the nature of the final data output that is required.

9.1 The optimal system

The most difficult step in the search for a suitable system is to define which criteria must be met. Its intended use, the questions which are to be asked, and the requirements of the survey should all feed into the criteria. A package for the scientific investigation of calls, for example, will have different system requirements from those needed for a comprehensive expert report for an environmental impact assessment. The recording of bat calls as a hobby or as part of publicity work will similarly have a whole different set of criteria. Operator experience and time and financial constraints, including follow-up costs, are also important factors to consider.

9.1.1 Intended use

The choice of suitable hardware depends on how it will be used. For mobile applications with manual detection and recording of calls, the system should include a heterodyne or frequency division detector, and possibly also the ability to store calls. It is also possible to run an automatic device in parallel with the manual system, so that no calls are missed. If there is to be regular stationary recording, the device must be able to record calls automatically and have a power supply for the required period of time. The most time consuming application is long-term monitoring over several weeks or months at one location. The availability of a power supply is an important factor, as automatic systems with high power consumption can generally be used only where there is a mains supply or where the batteries can easily be changed.

9.1.2 Acquisition and follow-up costs

It is essential to take into account not only the costs of purchasing a system, but also the costs of maintaining it in use. Fortunately, it is easy to gain a good overview of potential systems on the internet. As well as the actual hardware, it is important to evaluate and compare the obvious running costs, such as batteries or memory cards. There are also the hidden running costs of data analysis which are incurred with large quantities of data. One of these is the analysis of the recordings, which will be time consuming and therefore incur staff costs that will generally far exceed the cost of the equipment purchase.

Simple heterodyne detectors can be obtained for less than £100, while complex systems for long-term acoustic monitoring can cost more than £3,000. The running costs, for batteries or storage media, are negligible for the majority of normal usage, and there may be no costs at all for some systems. The picture with long-term monitoring looks very different, however. If a laptop is to be used in the field for long periods of time without a power supply, the cost of batteries quickly becomes prohibitive. It may well be that another system would be more cost-effective, even if the costs of purchase are higher. The use of a heterodyne system for long-term monitoring would cause staff costs to rocket.

9.1.3 Analysis software and further processing of the data

It is often conveniently forgotten that larger data quantities mean very long-winded processing. The recordings must be identified, and the results processed in various ways. If attention is not paid to recording quality when purchasing hardware, for example, it may be that the analysis can only be carried out manually. This could take several weeks and result in significant follow-up costs. The range of software systems is much more limited than the hardware, and it is worth finding out as much as possible from the producers. Some offer both a testing of the hardware as well as the analysis of selected recordings with their software. This will help with the selection of suitable software. It is also worth asking other users, which can be a quick and effective way of establishing whether the software is suitable for one's own applications.

9.2 Manual or automatic?

The most basic question is whether the system should be manual or automatic, or a combination of both. As discussed above, the system to choose depends very much on the intended use, but the aims and technical knowhow of the user also play an important part. A heterodyne detector will never be satisfactory unless it is often to be used outside at night. This type of detector is also not the right device for long-term monitoring or other time consuming applications. Of course, there is much to be said for its use in combination with an automatic system. Above all, a lot can be learnt about animal behaviour from observations carried out when using manual detectors. In the field of environmental education, it is absolutely the best device for bringing bats closer to an interested public. The echolocation calls can be easily converted to sounds audible to the human ear, enabling all the participants on a bat walk to enter into the secret world of bats.

Any environmental consultant choosing a heterodyne detector must be aware of the initial difficulties in identifying bats by this method, which relies on the human ear alone. If there is no training or experienced tutor available, it is possible to teach oneself with CDs of bat calls. However, it is difficult to work alone without help and advice, as mistakes are inevitable as a beginner. If the basic knowledge is incorrect, this will only lead to further errors being made down the line. Learning the skills of bat identification involves a significant number of evenings and nights spent out in the field. Even for experienced users of bat detectors, there will often be new situations when identification is still very tricky. This can be very frustrating when working alone. The same also applies to automatic identification systems, as there always has to be some manual checking. None of the current systems achieves more 60–80% accuracy in daily use, and in some locations or conditions the figure is even lower. Only if the calls are of the highest quality will the figure for accurate identification exceed 90%.

It is now generally accepted that long-term monitoring is indispensable for most significant planning applications. For this reason, serious bat surveys will need to involve the use of automatic systems, and environmental consultants will require the appropriate familiarisation with, and training in, these methods. It is also advisable to test-run systems before they are used for real applications. This will help spot possible defects which could lead to loss of data. By collecting test data with the new system, it is possible to practise evaluation and analysis and optimise it for the coming project. Ideally, one should have a set of data obtained before the system or device is acquired. It is then possible to test whether the system can satisfy the requirements of the survey report, not only technically but also in terms of analysis and compilation of the data. Realistically, however, this is rarely possible because of the lack of large datasets and, above all, time constraints.

9.3 Recording distance and amplitude

The issue of the recording range of the system is frequently raised, and this aspect is often seen as one of the quality criteria when choosing a detector. The operating distance of the microphone is, however, far less important than is generally realised. Assuming a reasonable basic sensitivity, other aspects such as reliability or quality of recording are far more significant. One of the reasons for this is simply that the bat itself and the weather conditions have a far greater effect on the distance range than the detector.

The recording range is dependent on the microphone, the amplification of the microphone signal and, particularly, the volume of the bat call. It is important to realise that signals can only be digitally processed if they have been recorded clearly by the microphone. If the signals are muffled by the system noise because the volume of the call was too low, they will not be recognised in the digital signal. There are some misconceptions regarding the recording ranges of microphones or detector systems. Some users assume a recording distance capability of between 150 and 250 m – but this is a crude overestimate, as will be seen in the following text.

When considering the recording range, it is essential to begin at the source of the sound, as it is the frequency and volume of this sound which will determine the recording parameters. Only then is it appropriate to factor in aspects such as the physics of sound propagation and the technical features of the microphone and the detector.

9.3.1 The theoretical aspects of recording range

Call volume

Bats exhibit considerable variability in the volume of their echolocation calls. As a rule, large bats call more loudly than small bats, and any one species will call more loudly in an open space than in a confined space. There are, of course, exceptions to this; for example, a bat may call quietly when it is homing in on close objects.

This behaviour can be observed close to wind turbines. Although the bat is in an open area, it begins to adapt its echolocation calls as it approaches the nacelle. Bats such as the barbastelle may adapt the volume to the hearing capacities of their prey, while others modify their call according to the presence or absence of competition from other bats. Bats can change the shape of the sound lobe or focus either the entire call or a single frequency segment. They can also affect the directional characteristics of a call by controlling the precise direction of the sound beam. These behavioural adaptations are, however, not dealt with in the following theoretical discussions.

One of the loudest sounds produced by a bat is the call of the noctule, which has been measured at 136 dB sound pressure level (SPL: see Section 14.1.2, *Sound pressure and sound pressure level*) 10 cm in front of the bat. This is roughly equivalent to the sound of an aircraft taking off. This is probably exceptional in the bat world, and a more normal value is around 130 dB. Other bats emit much quieter calls, such as *Plecotus* (long-eared bats) and some *Myotis* bats which typically are in the range of 90–100 dB SPL. Species such as the common pipistrelle reach 110–120 dB SPL.

Propagation

The SPL halves (which equates to a reduction of 6 dB) as the distance doubles. This is known as geometric attenuation (see Section 14.1.3, *Propagation of sound*). Thus, for the noctule above, 136 dB becomes 130 dB at 20 cm, and 124 dB at 40 cm. The spreading of the sound has a strong effect on the measurable SPL. While the SPL at the source of the call initially decreases rapidly, the rate of reduction becomes progressively less as the distance covered increases. From 100 dB at 6.4 m it will decrease to 94 dB at 12.8 m, representing a halving of the SPL. There will be a similar 50% reduction from 94 dB at 12.8 m to 88 dB at 25.6 m.

Attenuation due to atmospheric conditions (see Section 14.1.3) is an even stronger factor than geometric attenuation in limiting recording range over long distances.

Figure 9.1 shows how the strength of a 20 kHz sound source at 136 dB SPL, which is the maximum volume of a noctule, changes as distance increases. Three attenuation curves were chosen by way of example. The curve for an attenuation of 0.2 dB/m is very optimistic and would not be normal for natural conditions. The curve furthest to the left (with the square symbols) shows the trajectory for 0.5 dB/m, and is realistic for the frequency selected. The red horizontal line marks the SPL that can be detected by currently available microphones.

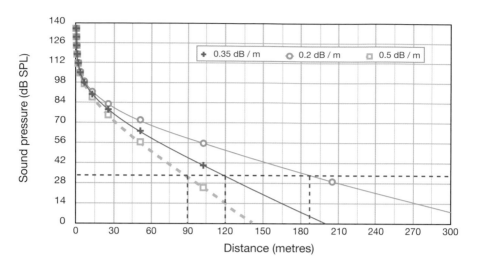

Figure 9.1 Sound pressure level (SPL) reduction at various distances as a result of atmospheric attenuation. The illustration here uses the maximum-volume call of the noctule by way of example (20 kHz, 136 dB SPL).

Recording distances

At best, the call of a noctule can be heard at no more than 150 m (see Figure 9.1). Realistically, 90–120 m is the maximum recording distance, even with the most sensitive microphone. Noctules probably only rarely call at sound pressure levels of 136 dB, and the microphone sensitivity must be assumed to be lower. With the system noise, that would be 36–40 dB. A maximum recording distance of between 50 and, at best, 120 m would then be possible.

For higher frequencies the maximum recording distance is lower still, since attenuation values can exceed 1 dB/m. Figure 9.2 shows the recording distances for quieter sound sources of 130 dB SPL at 20 kHz and 120 dB SPL at both 40 kHz and 50 kHz. It is easy to see the considerably reduced recording ranges at higher frequencies compared to lower frequencies.

The values shown lie in the upper third of the possible maximum recording distances. The values in Figure 9.2 will mostly not be achieved. As an estimate of the maximum range, the values can be seen as good enough.

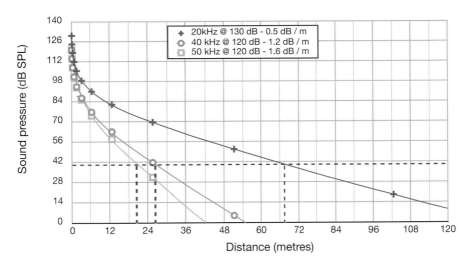

Figure 9.2   Sound pressure level at various distances, calculated at various frequencies with specific values for atmospheric attenuation.

9.3.2 Detection threshold

In Figures 9.1 and 9.2, a red line marked the minimum SPL at which a call would be recorded. This detection threshold is affected by the signal-to-noise ratio (SNR) of the microphone and the analogue circuit. The trigger algorithm will determine the minimum threshold volume that will set off the recording.

The distance over which a bat can be heard is thus dependent not only on the physics of sound, but also on the hardware used, particularly the microphone, and on the analogue and digital signal processing. Every circuit has its own noise, which is independent of the signal. This noise plays over the signal and can obscure low-volume calls. The background noise of the microphone is around 26–30 dB.

It is realistic to assume that the background noise of the entire system will be in the order of 40 dB. This means that the signal available for analysis is about 70–90 dB lower than the signal emitted by the bat. In Figure 9.2, this results in a red line at between 30 and 40 dB, in order to determine the recording distance. If there is additional disturbance from the surroundings, such as wind or mechanical noises, the SNR will be even lower.

Triggering

As well as the purely technical properties of the analogue hardware, the characteristics of the triggering system process need to be considered. As a rule, the devices do not record continuously, but are triggered when a particular event occurs. The trigger algorithm also has an SNR, which must be added to that of the hardware.

A greater recording distance is not necessarily achieved by so-called gain, the amplification of the analogue signal. Gain will increase not only the usable signal, but also the system noise in equal measure. Therefore, a signal which has already been swamped by noise will suffer the same fate with gain.

If the recording threshold is very close to the noise level, there will be very many more trigger failures. For example, the noise level at a wind turbine increases as the wind gets stronger, and this means that the number of recordings will progressively increase. Therefore, an increase in the maximum recording range, bringing with it a lowering of the recording threshold, will lead to a significant increase in the number of recordings. These recordings must then all be checked for bat calls. A large proportion will have to be manually checked, but this is only feasible if the number of calls is not too great.

9.3.3 Recording range in practice

The maximum detection distance is mostly lower than the values quoted above, because, for one thing, environmental factors will cause significant atmospheric attenuation. Furthermore, bats adapt their call volume to the particular surroundings or activity, such as hunting or feeding, and will call more quietly, at between 6 and 20 dB, if appropriate.

In such circumstances, it cannot be assumed that the maximum recording distance will apply. This would lead to an incorrect analysis of the data, as there would be an expectation of a much greater number of recordings. Far better, and more scientifically correct, is to provide minimum and maximum values. By way of example, the common pipistrelle would have a range of between 15 and 35 m. Without knowledge of the recording device used, however, it is not possible to specify the range any more precisely. This is because the amplification and the signal processing of the particular device are important in deciding the position of the red line in the graph (Figures 9.1 and 9.2). If no calibration of the microphones is carried out and no precise reference is specified for the recording threshold, it is better to err on the side of caution and underestimate the recording range. This contradicts what is stated in many books on the subject, which sometimes give very optimistic figures. The age of the microphone can also have a significant effect, as the sensitivity of these devices usually declines with use.

9.3.4 Recording three-dimensional spaces

The recording range of the detector system indicates only the potential distance at which a bat can be recorded. From this, it is then possible to derive the volume of the three-dimensional space which can be monitored. In the ideal case of a detector placed near the ground without any obstructions, the space is in the form of a hemisphere. If the microphone is placed in the open at a sufficient height, then a sphere can be assumed. Figure 9.3 shows the volumes of spheres for various detection distances. The recording distances may be affected, on the one hand, by the volume of the calls and, on the other hand, by the sensitivity of the microphone. If a microphone has a range of 40 m, then spatial volume at 20 m corresponds broadly to a microphone with 6 dB less sensitivity, that is, half the sensitivity.

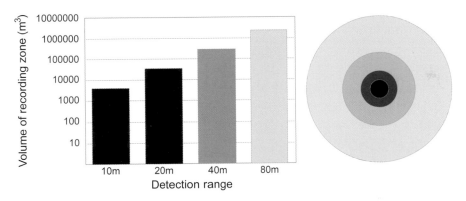

Figure 9.3   The volumes of zones used for recording are shown for various detection distances. As a rule of thumb, the differences between successive distances always represent a doubling of sensitivity.

The recording range is not necessarily the same in every direction. For one thing, the type of microphone will have a certain directional characteristic (see Section 14.2.2, *Directional characteristics*). The setup of the detector and the installation of the microphone also place further constraints on the directional sensitivity. For these reasons, the three-dimensional space that is being monitored will, in some circumstances, deviate considerably from the ideal form of a sphere (Figure 9.4).

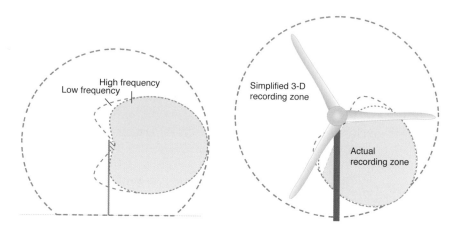

Figure 9.4 Illustration of the restriction of monitored space, brought about by the particular way a detector is installed, and here exemplified by two typical outdoor situations. On the left is a setup for a long-term monitoring project on a mast. On the right is a typical monitoring application of a detector installed in the nacelle of a wind turbine. The diagrams depict a simplified view of the recording zones.

The microphone can usually only be used when integrated into the detector. The way in which it is fitted can sometimes mean that the microphone is shielded partially or entirely from the sound coming from certain directions. This depends on the wavelength of the sound (refraction) and therefore also on the frequency. Shorter wavelengths will be more prone to this problem. This aspect is often not taken into account in the processing of acoustic data, and, for the sake of simplicity, the assumption is made that sound reception is uniform from all directions. This is an acceptable strategy, as empirical measurement of the actual space being monitored is not always feasible. This is illustrated in Figure 9.4 with two typical recording situations. In order to emphasise the complexity of the process, Figure 9.4 is laid out very precisely with reference to the space being monitored. The effects of the deflection of sound waves by the obstacles cannot be explained in simple physics. In the case of small obstacles (as in the left-hand diagram), lower-frequency sound coming from a source behind the object will presumably still be recorded, albeit at a much-reduced range. It must also be borne in mind that these are two-dimensional cross-sections across a three-dimensional space.

In many manuals and reports, the recorded activity is recalculated to take account of the relationship between the assumed recording distance and the rotor diameter. Figure 9.4 clearly shows that this kind of correction of results may be risky, as only one part of the potential recording space is being monitored, and nothing definitive can be said about the remaining part. Here, the number of bats calling is unknown, since a uniform distribution over the study space is unlikely. Therefore, the value for the activity level cannot be corrected on the basis of the assumed volume of the recording area.

9.4 Triggering systems

The significance of call recognition or triggering of the recording for the assessment of the recording range of a detector has already been mentioned. As well as the actual sensitivity of the microphone and amplification, the production of recordings also depends on the recognition of a positive signal. The exact workings of a trigger function are not known in detail for any of the detectors, even though it may be one of the possible unique selling points. The reliability and objectivity of a trigger mechanism is important for the evaluation of a device. If a bat call triggers a recording once, it must do so for repetitions of the same call. This means that the trigger algorithm must set off a recording even if, for example, there is a change in background noises, provided the latter do not superimpose on the call itself.

Devices that have an adaptive recording threshold, instead of a fixed one, must be treated with caution. A fixed recording threshold means that, irrespective of the signal input level, this threshold has to be exceeded in order to trigger the device. It is generally a threshold which is relative to the maximum recording level. Both the Avisoft USG and the batcorder have this type of fixed threshold. Ideally the system is calibrated for sensitivity, as it is only in this way that a fixed threshold can be implemented meaningfully. According to its specifications, the Pettersson D500 also has a fixed threshold. This can only be set at three levels, however, and no precise details are provided on the actual trigger threshold levels. Moreover, the device is not calibrated. The Batlogger has single triggering modes (*Period, SD-Trigger*), which presumably have a fixed threshold, but this is not clear from the settings or documentation. Neither the microphone nor the device is calibrated. According to the manufacturer, both triggers are less sensitive to species with low-frequency calls. That means that, under certain circumstances, the system will not be triggered. There is therefore a possibility that species from the genera *Nyctalus, Eptesicus, Vespertilio, Tadarida* and *Plecotus* will be missed.

The problem with a variable or adaptive recording threshold is that the triggering of the recording is unreliable or not always reproducible. The detector changes the recording threshold according to the level of ambient or background noise. A sound will trigger the device once it exceeds the set value. If, for example, bush-crickets are active during the summer, or the location is 'loud' because of other noise sources, any bat calls will have to be correspondingly louder (Figure 9.5). Adaptive thresholds are used in the SMxBat devices from Wildlife Acoustics as well as in the Batlogger models (*crest, crest-advanced triggers*).

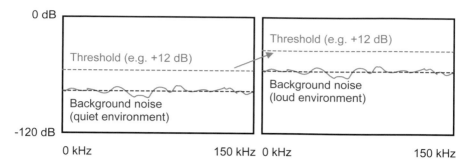

Figure 9.5 When using an adaptive trigger threshold, the background noise affects the triggering of the recording. If the ambient noise increases, a trigger signal will automatically have to be louder.

Some models allow the activation of a 'dead time' after completion of a recording, in order to reduce the number of recordings from single individuals. The option exists with the Pettersson D500 (*minimum time interval to the next recording*) and with the Batlogger (*Trigger-Ignore*). These functions have to be deactivated when using the devices for projects carried out for the purposes of impact assessment reports or research, since particular species could otherwise be missed.

In the trigger settings for all devices, there is the option for frequency filtering, in which either high pass or low pass settings can be selected. With the high pass, only frequencies above a cut-off level will be included, whereas with the low pass, only frequencies below a cut-off level will be included. This facility can only be used when certain frequencies genuinely do not need to be monitored. Otherwise, some species will be lost by the incorrect use of a filter.

9.5 Recording quality and identification

When embarking on acoustic recording of bats for the first time, there is a huge range of knowledge which has to be acquired in order to understand the methodology and to be able to use it competently. In addition, there will be the pressure of deadlines, and there is usually barely enough time to acquaint oneself with the systems and techniques. It is often difficult to appreciate the consequences of the technologies chosen for the analysis and interpretation of data. This can then result in the preparation of costings and tenders which are not necessarily viable.

The greatest problems often only surface later in the project during the analysis phase, when the quality of the recordings turns out to be inadequate to carry out accurate identification. For long-term monitoring, automatic identification is a must. The quality of the recorded calls has a very considerable impact on (automatic) measurement of parameters and identification.

If automatic methods are not feasible because the calls are faint or have too much background noise, it will be necessary to resort to manual identification. As has already been discussed in previous chapters, this represents an enormous time commitment which should not be underestimated. The right choice of device and its sensible application are therefore critically important. All available devices can more or less produce good recordings, but what is important is to choose the correct setup and settings. Experience over the years has shown that some systems, without an optimal setup, too often produce calls that are unsuitable for reliable automatic processing.

9.5.1 Setup and echoes

Every device has strengths and weaknesses in the quality of its recordings. In particular, a poorly positioned microphone can lead to inferior recordings as a result of echoes (see Section 8.2.2, *Multidirectional spread and reflections*, and Figure 8.7). As well as reflections off the device itself, deeply embedded microphones also present problems. Dispersion effects can develop at the edge of the hole in the casing, and this can produce noise in the recording similar to echoes. Many devices will therefore need to be modified to obtain echo-free recordings.

When setting up devices in the field, it is essential to ensure an echo-free environment. That means that the microphone must not be located behind a screen of leaves or inside a nest box. The various rain-protection fixtures available have also proved to cause echoes or overlapping. This means that automatic analysis is no longer feasible, and that even manual processing of calls is difficult (see Section 9.6, *Weatherproofing*).

9.5.2 Recording location

The location has implications for the quality of the recording and the reliability of the identification, as it will have its own particular echoes and interference. Even atypical calls such as those in the vicinity of roosts can also have an impact on the recording and its analysis.

Even the best hardware cannot produce clean recordings when set up in or near highly reflective structures, such as underpasses or thick foliage. All the recordings will contain interference to a greater or lesser extent, and will be difficult to classify (Figure 9.6). Better results will be achieved only by intelligent adaptation to the particular conditions of the location. As a rule, structures that will clearly cause interference can be easily spotted on the first inspection visit. Some of the causes of echoes, however, may not be immediately obvious, as one of the factors is the direction of flight of the bats. Finally, other noise sources nearby may have a negative effect on the detector.

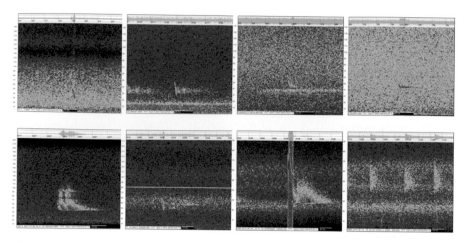

Figure 9.6 Some devices often produce noisy recordings because of the way the microphone is fitted, or because of other technical conditions. These sonograms show calls that may prove difficult or impossible to analyse.

Atypical calls which the recognition software does not or cannot know are a further source of error. As a rule, this applies particularly in and around roosts. These types of errors are usually independent of the system used, although it has to be said that some systems are more prone than others to these problems.

At all events, manual identification of the recordings can still provide useful results, so that recording at these types of location will not inevitably be a wasted effort. However, the time input required should not be underestimated. For this reason, it is best only to investigate this type of location using acoustic methods if useful data cannot be obtained in another way.

9.5.3 Sensitivity and distance

The device settings have a significant impact on the recording quality and hence on the identification. Generally, devices are set to very high sensitivity, as the aim is to record as many bats as possible. This leads to the recording of very-low-amplitude calls, which arrive at the microphone in a piecemeal fashion. While the recording in general may be good enough for reliable species identification, these recordings will have to be analysed manually.

Higher sensitivity can always be achieved by a higher amplification of the signal. This is a thoroughly sensible tactic in some instances, although it is important to realise that the system noise will also be amplified. This means that the signal-to-noise ratio is not effectively improved, so call segments which are near or within the interference band will not be lifted out from this zone. Sensitivity over longer distances does not always mean better monitoring or more data. It is much more effective to limit distance range in order to maximise good-quality calls and minimise poor-quality ones.

In this way background noise, which makes data analysis very difficult, will be kept to a minimum. High sensitivity settings will also lead to a large number of extraneous recordings such as crickets.

Conversely, this does not mean that sensitivity should be set to the minimum. The correct settings will be determined rather by the nature of the investigation and the technical solution adopted. It may be that the device employed is able, even at high sensitivity settings, to produce clear recordings which can be analysed for the purposes of identification. The story will be very different, of course, if the quality of the recordings is poor.

It is, in general, inadvisable to opt for amplification unless the microphone is calibrated. Without knowledge of the degree of sensitivity of the microphone, the selection of amplification settings on the device is purely a matter of taste. The most commonly used microphone from the Knowles FG range has an output variation of ±3 dB to ±6 dB. This means that one microphone can be more than twice as sensitive as another if this is not corrected by the manufacturer. The implication of all this is that the differences in microphone sensitivity and range make it difficult to compare data acquired from different setups.

The sensitivity of the system that will actually be required will depend largely on the nature of the investigation to be undertaken. As has been discussed in Section 2.2, *Data quality and its implications*, there will be a choice between qualitative and quantitative data. It is this, and the range of species studied, that will finally determine the optimal sensitivity needed.

9.6 Weatherproofing

A recurrent theme is the need to protect microphones from the weather, particularly rain. If long-term monitoring is being carried out over several months in inaccessible locations, the microphone needs to be sheltered as much as possible, in order to avoid a deterioration in its sensitivity. There are many solutions proposed for microphone protection. Unfortunately, most of these have undesirable side effects.

A microphone must be set up so that it can receive sound effectively, and so must have direct contact with the environment in which the recordings will be made. As a result, it will be subjected to all sorts of environmental influences, not the least of which is rain. Damp is a problem for any electronic device in the long term. The sensitivity of microphones to damp conditions will vary greatly and will depend on their construction and the way they work. Large membrane microphones are generally more sensitive to damp than small electret ones. The former, however, are more sensitive to sound and have a better frequency response.

There are various solutions available to improve weatherproofing, and these are presented briefly below. The list makes no claims to be exhaustive.

The pros and cons of the systems that are not mentioned here can probably be derived from the descriptions of other systems.

9.6.1 Sound deflection

A solution to the problem of protection from the weather was first developed for the Anabat system. It is relatively well known and is in frequent use. The general principle is based on sound deflection off a panel (Figure 9.7a).

The microphone is encased in a tube and is directed downwards so that it cannot be damaged by rain. Underneath the tube, at a suitable distance, a plexiglass plate is attached in either a horizontal or a slightly sloping plane. This reflects sound coming from above up into the tube with the microphone, which is otherwise largely shielded. While the microphone is now well protected, the call is subject to significant modifications, since echoes are produced by the plate and the inside of the tube. Automatic analysis becomes almost impossible, and even manual identification is rendered more difficult. With the Anabat system, this is less marked because of the reduced data of the zero-crossing analysis.

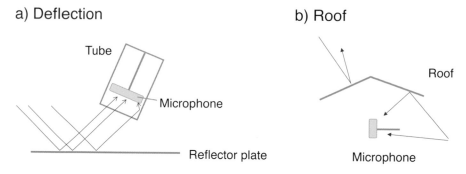

a) Deflection

Tube

Microphone

Reflector plate

b) Roof

Roof

Microphone

Figure 9.7   (a) Deflection of sound by means of a plexiglass plate. The microphone sits in a tube, protected from the rain. This setup can lead to significant interference in the recording. (b) Protection by means of a roof. This can similarly lead to echo problems and part of the calls being suppressed.

9.6.2 Roof over the microphone

A roof over the microphone is one of the solutions which is sometimes adopted to afford protection from the rain. This does shield the microphone from some of the sound coming from above the roof (Figure 9.7b), and the roof itself will create echoes which will be superimposed over the call (interference). As with the solution discussed in Section 9.6.1, these deflections will render the analysis much more difficult. This solution is far from optimal and should not be used.

10 Interpretation of the results

Once the acoustic data have been collected, the next issue is how to interpret them. In the case of species groups other than bats, it is often possible to establish absolute numbers and thus population sizes. Indeed, many ecological studies are based on the numbers of individuals.

Since it is not possible to distinguish between individuals in acoustic recording, quantification of activity in the form of the absolute number of bats is not feasible. Therefore, measurements of relative activity must be used instead, and here there are various approaches to analysing the data collected. By choosing the most appropriate recording technology, qualitative and semi-quantitative statistics can be obtained. Analyses of this type are often needed in landscape planning, when changes need to be assessed.

10.1 General problems

Recordings of bat activity are usually gathered in order to undertake an analysis of the data. In addition to the indices described in detail later in this chapter, bat behaviour and the general limitations of the technology present various problems, and these cannot always be solved by strategies such as increasing the number of studies or making greater efforts in the field.

10.1.1 False negatives

One of the general problems encountered in ecological investigation is the issue of false negatives. If the presence of a species is not confirmed by the methodology applied, this does not necessarily mean that it is absent from the habitat. Each method has a specific effectiveness with regard to proving the presence of a species. The behaviour and the ecology of the species studied are important factors here. Nobody, for example, would use an ultrasound detector to search for bats during the day, when the probability of encountering them would be close to zero.

In order to prove the presence of a species acoustically, it must be heard on the detector or recorded digitally. The bat must be present within the detection range, so it is clear that the effectiveness of the method is proportional to the sensitivity of the detector. The recording must be of high enough quality to allow identification (see also Section 2.2, *Data quality and its implications*).

Species that are very common, call loudly and are mobile hunters in small spaces are generally easier to identify. Loud calls carry over long distances, and frequent occurrence and high mobility increase the likelihood of the bat passing the microphone. Each of these factors on its own will not, however, increase ease of identification. If a bat is common, but has faint calls, then it is easily missed. The genus *Plecotus* is a good example of this, as it mostly forages for prey using passive echolocation (i.e. by listening to prey sounds). It is thus easily missed, although it may occur in many different locations. Similarly, a loud, but rare, species may easily be missed. It is also difficult to detect a loud species that is highly mobile and ranges over a large area. These kinds of species generally occur in low densities and are as difficult to detect as rare bats. Highly mobile species, such as the parti-coloured bat, are probably unrecorded in many parts of Germany simply because of their rarity. However, another factor contributing to the lack of records for this species must be the difficulty in identifying it by its call. But even the occurrence of a commoner species such as the noctule is difficult to prove because of the long distances it travels from its roosts. On the other hand, recording calls close to the roost site or to food-rich habitat such as an area of water makes it easier to detect the presence of a species (Bruckner 2015).

It is easy to make a list of species that are difficult to record using acoustic techniques on account of two of the factors mentioned above, namely loudness of call and difficulty of identification. Even if a positive record for these species is obtained, it may not be possible to investigate them further because of the difficulty of recording them. These species are:

- Long-eared bats (*Plecotus*) – all species
- Bechstein's bat (*Myotis bechsteinii*), Geoffroy's bat (*M. emarginatus*) and possibly greater mouse-eared bat (*M. myotis*), Brandt's bat (*M. brandtii*), Whiskered bat (*M. mystacinus*), Daubenton's bat (*M. daubentonii*) and pond bat (*M. dasycneme*)
- Horseshoe bats (*Rhinolophus*) – all species
- Parti-coloured bat (*Vespertilio murinus*)

Thus, the lack of a record for a species should not be equated with its absence from the study area. Rather the question needs to be asked whether the recording methods were good enough to produce a positive record. The example of a long-term recording study in Section 2.7, *Assessing biodiversity*, showed that the average number of species on any one night tends to be quite low. This example is only anecdotal, but other comparable studies in open environments show similar trends. Figure 10.1 shows a similar distribution of recordings each night. On many nights there are only a few records, and it is only on occasional single nights that higher numbers are achieved. Long-term recording will usually show this pattern of distribution.

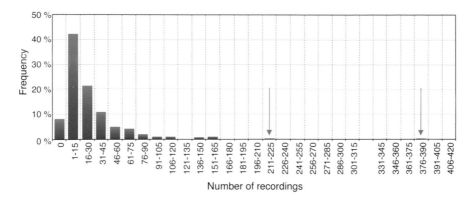

Figure 10.1   These data were recorded during a long-term monitoring study of open countryside at four locations using the same recording methods for 773 nights. The average number of recordings per night was 29. The arrows show sparsely populated data ranges

It is very difficult to say precisely how many recordings are necessary to confirm a negative result. It depends on factors such as the range of species, the detector settings and the precise location of the detector. Some species, such as the common pipistrelle, are very easily recorded and documented after only a few days. However, even a common pipistrelle will venture over the middle of an agricultural landscape without hedges and other structures on rare occasions during migration or because of extreme weather or acute shortages of insects. If recordings are made at the wrong time, then this event will not be captured. It is therefore advisable never to state that a species is definitely absent from a given location. Long-term monitoring, which is now standard for such projects as wind farm planning, produces the most reliable results for the very reason that is carried out over a long duration. Walking transects over a few nights, in contrast, just does not produce such authoritative results. From Figure 10.1, it is obvious that in almost 10% of all nights no recordings were triggered. Approximately half of all nights produced 15 or fewer recordings.

10.1.2 Average values

The data obtained for each species are used for calculating averages for reports. There will be statements such as *two recordings on average* or *0.5 seconds activity per night on average*. As a rule, the arithmetic mean is calculated over all the nights of recording. This average indicates how often or how seldom activity was recorded. It is mostly interpreted in this way, even if, from a statistical point of view, such a statement is not necessarily strictly correct.

When calculating and interpreting averages, the distribution of the data must always be taken into account. If the data have a normal distribution, the arithmetic mean can be used for the purposes of analysis. If that is not the case, however, the distribution and, where relevant, the median value should be described in detail. If the distribution is skewed, the relevant statistical properties will change. Data gained from acoustic recordings tend not to be normally distributed. They can be regarded as count events and represented as a Poisson distribution. If appropriate, a log-normal distribution would also be a good approximation, as illustrated with an example dataset in Figure 10.2. Normally, acoustic data show a high proportion of low activity with outliers of extremely high value. This makes the application of statistical tools very tricky.

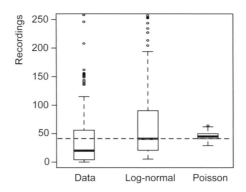

Figure 10.2   The actual distribution (left) of acoustic data obtained over 202 nights compared to a log-normal distribution and a Poisson distribution with the same expected value (dotted line)

The species being studied can also have an effect on the distribution, as can the location and the ecology of the species. Figure 10.3 shows the *minutes with activity* obtained at two locations for four species. The presentation of the data is enhanced by the log transformation.

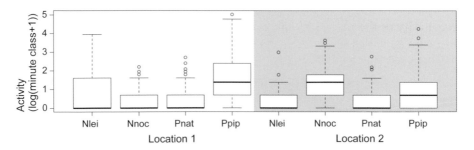

Figure 10.3   Data from 189 nights of two long-term recordings, separated into species. They are plotted against logarithmic activity in minute classes. Nlei, Leisler's bat; Nnoc, noctule; Pnat, Nathusius' pipistrelle; Ppip, common pipistrelle.

An average value always represents an estimate, because it is not calculated for the totality of the data, but rather is based on a sample. How the average is to be interpreted depends on the distribution of the data. If this does not have a normal distribution, some of the usual statistics will not be available.

The negative results, which have zero values, present a further problem for the calculation of the average. It must be decided whether these are genuine zero values to be included in the calculation, or whether they are so-called NA, not available, missing values. The latter should not be included in a calculation of the average. As was described in Section 10.1.1, this distinction is by no means trivial. If data are missing, for example because a species is not always present owing to its migratory behaviour, this must be taken into account in the analysis of the averages when the description of the activity concerns an endangered species. In Figure 10.3, the Nathusius' pipistrelle (Pnat) is just such a case. It is present in the study area during the migration period, but is otherwise almost totally absent. As a result, the distribution is skewed towards zero. Outliers then occur on nights of high activity during the migration period.

As a rule, average values have no meaning if the survey consists only of a few transect walks. If bat activity is investigated over ten transect walks during the year, individual outliers will strongly affect the results. Careful consideration must be given as to whether calculation of an average will be meaningful or whether, because of the small amount of data, it will be a poor estimate.

10.1.3 Different species are not directly comparable

Differences in foraging techniques and echolocation methods prevent a direct comparison of results for different species. The volume of the calls, the distance range (which is dependent on the frequency – see Figure 10.4), and the call gaps all have a significant impact on the data obtained. Furthermore, foraging behaviour will determine the frequency of the recordings. The sensitivity of the methods, and hence the quality and comparability of the recordings, varies from species to species.

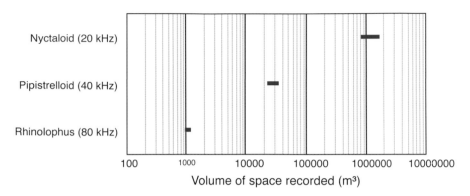

Figure 10.4   The recording spaces, measured as a volume, for three genera show clear differences. The volumes are based on calculations by Jens Koblitz which take into account the spread of the sound. Allowance had to be made for relative air humidity values ranging from 38% to 86% and temperatures from 17.5 °C to 27.9 °C (1 July 2015, measured data for Konstanz). A call volume of 115 dB SPL at a distance of 1 m was chosen. The volume (m³) was calculated as a hemisphere.

For this reason, all activity indices must relate only to a single species when comparing activity between locations or individual nights. As a rule, it is not possible to make direct comparisons between species. It may be possible to consider a number of species together, by drawing on derived values such as the percentage use by one species of particular locations. However, such comparisons are really only relevant in scientific studies. In environmental impact studies, species or groups of species are usually treated separately from one another.

10.1.4 Analysis by category

In practice, a distinction is often made between categories of activity: none, low, medium and high. Individual bat workers have developed their own scales, by means of which the number of recordings per hour or per night are analysed. No distinction is made between species, and the same scale is used for all species. The analysis must take account of variables such as foraging methods, call volume and mobility of each species. The type of planned development will determine which scale is most suitable. Moreover, the value ranges will be based on the empirical values determined by the bat workers themselves. To what extent these are adequate for the particular application cannot be determined by reading the reports. As every bat worker creates their own activity categories, it is difficult to make meaningful comparisons between studies.

As a rule, an activity analysis should not be based on the absolute number values, but should take account of activity patterns within a defined space or time. If, for example, 500 *activities* are recorded for the noctule over one year, that may be very little when converted to the rate per day. However, if the same number was recorded over a few days, this figure assumes a greater significance in the context of the project.

10.1.5 Mobility and opportunism

The effect of the recording location on the data acquired is considered elsewhere, particularly in Section 10.2.1 (Figure 10.7). In general, bats are highly mobile and may often behave very opportunistically. That means that randomly occurring concentrations of insects will quickly be spotted and exploited by bats in the area. This is particularly true of bats that fly in large open environments away from landscape structures such as hedges and tree lines. This has implications when evaluating activity, as the data obtained may be skewed by recording at the wrong location. Such erratic results will become statistically insignificant, however, if large quantities of data are collected over a longer time period, ideally several weeks. For this reason, recordings carried out over single nights will not necessarily produce a true picture of the situation.

Mobility has another implication for the analysis of activity. As the roost and the feeding area may be in separate locations, bats will have to travel longish distances between them. In some species, there will be easily recognisable commuting paths which are taken by many, though not all, of the bats in the roost. Some bats will take their own individual routes to the feeding areas. This will vary according to the nature of the roost and the time of year.

If an activity index uses the number of feeding buzzes, for example, these commuting paths may not always be assigned their proper significance in the figures. The commuting path may be very important for certain individuals on their way to the foraging area, but this will not be recorded or acknowledged in the statistics, as the feeding buzzes recorded will be zero or small in number.

Mobility and opportunistic behaviour should always be considered in the data analysis. Neither can be measured, but they are ever-present factors. It is therefore important not to place too much emphasis on an index which may sometimes ignore important locations. This may lead to decisions being taken which are prejudicial to the bat population in the area. Longer periods of recording or long-term recording will help to avoid this. Large quantities of data will aid the proper understanding of these types of locations by taking account of time or regular usage patterns.

10.2 Quantification of activity – using identical recording systems

This section describes some of the measures of activity which can be employed, and which all rely heavily on the technology used. The following indices only go some way towards allowing comparisons of data that are not affected by the technology used. Certain recording techniques are generally unsuited to some of the indices described. Some of the following indices vary greatly with respect to their suitability as a measure of activity.

10.2.1 The number of recordings

The number of recordings may be a simple measure of the frequency of 'bat passes'. For simplicity's sake, the number of recordings is often equated with the number of bat passes, although this is not strictly speaking correct. The number of recordings is affected both by the technology and by the behaviour of the bats.

Some of the devices available only allow the user to set a constant recording length. The number of recordings will depend on the length of recording set. In other words, the number of recordings is inversely correlated to the length of recording selected. The longer the time setting that is chosen, the fewer, as a rule, will be the number of recordings. The results of different recording systems will vary considerably, so it is impossible to interpret this number in a report or assessment without knowing the recording length selected.

Other systems trigger a recording as soon as a bat call is detected and stop when, for a predefined time span, no more calls are detected. The minimum necessary pause between two sequences (defined by the post-trigger setting on the device) determines the number of recordings. With shorter defined time spans, more recordings will be made.

If the data are to be used in statistical analyses, care must be taken when working with recording devices, even if they are of the same type or construction. It is important to ensure that the recordings are made with identical settings. Otherwise, the activity figures are not suitable for comparison. Different settings will produce different results. Meaningful comparisons across different technologies based on the number of recordings are therefore not feasible.

Figure 10.5 shows the comparison between a system with recording lengths of 5 and 60 seconds and a variable system with a maximum call pause of 600 milliseconds. The longer the fixed recording duration, the greater the statistical spread of the result in comparison to call-triggered recordings and short recording durations. In order to compare data which have been captured in many different ways, it is necessary for this proportion to have a fixed value. In the example, it fluctuates between 40% and 90% of the actual activity. Thus, with a system of this type, the number of recordings is a not suitable basis for a comparative discussion of activity.

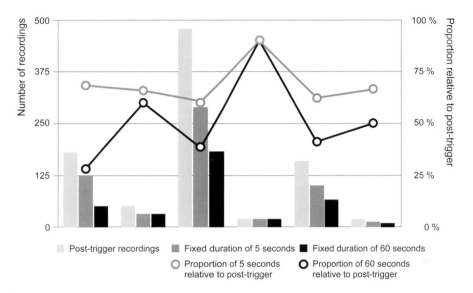

Figure 10.5 Numbers of recordings obtained with a device of variable recording length (maximum call pause of 600 ms) in comparison to the number of recordings with a fixed recording length. The number of recordings for the fixed time lengths, when expressed as a percentage of the number of recordings for the variable/post-trigger recordings, demonstrates the greater statistical spread with a longer recording length.

Even when using the same device, however, different settings will result in different numbers of recordings. The batcorder can be used as an example to illustrate this, showing the effect of adjusting the post-trigger value. This user-chosen value gives the duration of the pause in a call before a new recording is started.

Figure 10.6 shows an example of a Leisler's bat with call intervals of 300 ms, flying past a location ten times in a night, calling seven times on each occasion. With a post-trigger of 200 ms, the result will be 70 recordings, with one recording per call, since the call intervals are longer than the post-trigger. At a second location, the Leisler's behaves identically, but the post-trigger of the device is set to 400 ms. In this case, a total of ten recordings will be made. As a consequence of this, an assessment at the first location will be rated seven times better if the number of recordings is taken as the criterion.

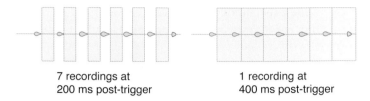

7 recordings at
200 ms post-trigger

1 recording at
400 ms post-trigger

Figure 10.6 Effect of post-trigger settings of 200 ms and 400 ms on the number of recordings in a call sequence of the Leisler's bat. Seven calls in a sequence result in seven recordings at 200 ms compared with only one recording with the longer post-trigger setting.

If all data are recorded in an identical way, this is not a problem. But if data are recorded using different settings, such as different recording durations, the results will be incorrect. For a major impact assessment study of wind turbines, for example, if post-trigger values are set too high, this can result in a significant underestimation of the number of bats affected or killed.

Furthermore, the behaviour of the bats and the precise location of the recording device also influence the numbers of recordings. For example, a bat may forage around the recording location, but its flight movements may take it in and out of the recording range of the device. This will produce varying numbers of recordings and suggest varying levels of activity. Similarly, the location of the device will determine the number of recordings. Figure 10.7 illustrates this with a simple example. If an assessment is based simply on the number of recordings, the activity appears to be five times greater at one location than at the other, because the bat only foraged within range for a short time.

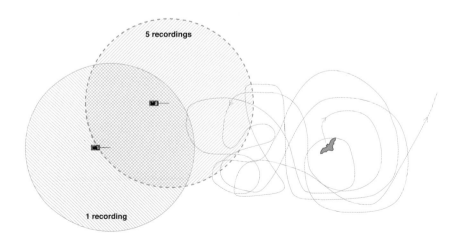

Figure 10.7 The number of recordings obtained by two devices with identical settings will diverge simply because of their different locations.

For all these reasons, care must be taken when using the number of recordings as the basis for comparing activity levels. Nonetheless, it is frequently used, no doubt because it is easy to read this figure on the display in the field, and no calculations are necessary. If used properly, it is a useful guide to the level of activity, and, as a rule, high numbers of recordings result from a high level of activity.

The converse may also apply, namely that few recordings almost certainly mean little activity. Modern programs allow the output of other activity information without the need for time consuming manual calculations. In summary, then, the number of recordings may be used as an indication of the level of bat activity, but it should not be used in comparisons of different sets of data.

The number of recordings is a useful measure when the bats are present in the vicinity of the recording location. A good example of this is at narrow points on a flight path.

10.2.2 The number of passes

The number of passes may be used as a good measure of bat activity. A heterodyne detector can be used in the field for this purpose. One pass is counted for each bat which can be audibly separated. It is often the case that the bat flies in and out of the detection range, which makes it difficult to distinguish between individuals, and a single bat may account for several passes. Nevertheless, it is sometimes possible in the field to determine whether a different bat is being heard, such as one passing along a flight corridor, or if it is the same animal foraging for insects. All in all, it is a very subjective process, and everyone will come up with different assessments.

A more objective count can be achieved by introducing a short pause between each new pass. However, the application of this becomes progressively more difficult as the length of the pause increases. With longer pauses, it is more accurate to talk not of the number of passes, but of activity per unit of time. It is therefore better to use short pauses in order for them still to qualify as passes. The pause should be technically and manually easy to measure in the field. A pause of one second is eminently suitable in this respect. Flight paths can be surveyed in this way, since, as a rule, there will be one pass for each bat flying through. Bats that are foraging will perhaps produce several passes, but at least the important role of the location as a foraging area will be well illustrated. Ideally, totals for each species should be counted separately.

In practice, the counting of passes using a hand-held detector is not always easy – especially, for example, when there is a high level of activity and possibly many species. It is not necessary to stand outside with a bat detector, of course, as there are automatic recording systems which can perform this task. Automatic systems have many different ways of controlling the recording. Some devices start and stop the recording according to the presence of bats. These sorts of device count passes more or less according to the above definition. Other systems, once triggered, simply record for a fixed period of time. This can lead to one pass being broken down and spread over several recordings. Equally, several passes, as defined above, may feature in one recording. In order to calculate the number of passes, therefore, it will be necessary to check manually for pauses. It is common to find the number of passes featuring in a report, but all too often there is no indication of how a pass is defined.

Since the number of passes does not usually differ significantly from the number of recordings (Section 10.2.1), all the issues relating to comparability of data also apply to the number of passes.

10.2.3 The number of seconds

A good measure of the intensity of usage of a location by bats is the amount of time they spend there. This means how many seconds or minutes in total a species or genus spends there each night. In comparison to the number of recordings, the time spent may be a better indication of the intensity of bat activity in a location. Longer periods of presence will thus receive a greater weighting than a simple counting of the number of recordings.

With some recording systems, the value must be laboriously extracted by means of a manual analysis, since, once they have been triggered, these systems simply record for a certain fixed time. For this reason, this measure is rarely used. Other devices start and stop according to the presence or absence of calling bats. This method allows the activity to be extracted in the form of length of stay in seconds.

The length of stay indicates how much a location has been used by a particular species, and is a good measure of activity. However, it is not possible to tell whether the recording is one bat using a location for a long time, or several using it, each for a short time. In contrast to the index *number of recordings*, the length of the sequences is evaluated and weighted accordingly. From the number of seconds, it is easily possible to derive a relative density. This will then serve as a measure of the preference for a particular location. By standardising the number of seconds to the length of the night or the duration of recording, a relative density can be established, and expressed in terms of *seconds of activity per hour of recording/night*. Using an appropriate standardisation, data can be compared, if the length of the night or the recording period varies.

The disadvantage of duration of stay as a measure of activity is that as bats adapt their call volume to the surroundings the recorded length of stay changes. With louder calls, it is mostly the longer sequences that are recorded. All species adapt their call volume dynamically to the surroundings. Thus, in locations with higher vegetation or structural density (clutter), it is to be expected that calls will be quieter, and the length of stay recorded will be shorter. Equally, no comparison of recordings made by different devices or with different settings is possible. Technology has an effect on the length of recording and distorts the results, just as it does when the number of recordings is used as a measure of activity.

10.2.4 The number of calls

In some articles, the number of calls is used as a measure of activity. There are, however, technical and methodological problems with this. The calls can scarcely be counted in the field. Furthermore, even when the calls are on a computer, an automatic solution is needed to determine the number. If calls are counted manually, there must be clear criteria for defining a valid call. This would include a minimum length and the call types such as social, echolocation and feeding buzz calls. During the counting process, a manual analysis of the calls may also allow a distinction to be made between bats that are passing through and those that are foraging. This assumes that it is possible to differentiate between these two activities by the bats' calls. The foraging technique of some bat species may not allow that.

Since bats adapt their calls flexibly to the environment and the task in hand, this measure is not always applicable. Thus, the call rate is often raised with an increase in vegetation or in cluttered environments. The noctule bat, which echolocates in open environments at a rate of two calls per second, will emit five or more calls per second when flying along a woodland edge. Other species, too, adapt the volume of their calls to the habitat. Long-eared bats often use very loud calls when foraging over meadows, but emit very faint calls when in woodland. Loud calls record better, and higher activity rates will therefore be detected at a location where the calls are louder. Closed habitat will produce comparatively lower activity rates, since the quiet calls will only rarely be recorded. The use of different devices or different sensitivity settings on any one device will also inevitably result in differences in the number of calls registered.

The number of calls is thus not really suitable as a meaningful index of activity. On the one hand, there are the problems mentioned of counting accurately and, on the other hand, the problems of the variable echolocation systems employed by bats. In this context, it must be mentioned that, when people talk about the number of calls, they often mean the number of recordings. The two are often incorrectly equated. This will only lead to confusion.

10.3 Quantification of activity across different technologies

Another way to quantify bat activity is to measure activity in specified intervals of time. This measure is suitable for comparisons of recordings made with different technologies or settings. The prerequisite for a comparison of results from different technologies, however, is that they must overall have approximately similar sensitivity. If there are significant divergences between devices, then this index may also only be usable under certain circumstances.

In the case of manual recording, activity is often portrayed as being related to the length of time for which that activity is recorded (point stop transects). This means that all passes occurring within the fixed time stopping period are considered an *activity*. This will then require a simple calculation to generate the index. Critical for the meaningful application of this technique is the duration of the time intervals, that is to say, for how many seconds or minutes activity is measured. A very short time interval will mean that, although the analysis is complicated, the result will correspond roughly to the number of recordings or passes. With a longer duration, the result is imprecise and less meaningful, comparable to counting the number of recordings taken over a long, fixed period.

Meaningful intervals of time depend to a great extent on the nature of the study. A feasible reduction of the data, which will not lead to too great a loss in accuracy, is achieved by using time intervals of between 10 seconds and 10 minutes in length. The optimal value is determined by the precise question to be addressed and the recording and analysis methods employed.

A good compromise is to use intervals of one minute in duration, which will capture short foraging flights and be easy to evaluate. Bats recorded which are clearly separated by time can be counted individually. Using a time interval such as one minute, which is adapted to the behaviour of the bat, it is easy to calculate an index of activity. The one-minute interval can be effectively used in a database or spreadsheet, simply by eliminating the seconds from the calculation. For smaller and larger classes it is more difficult, and it is often necessary to use complex macros or functions.

Even for a comparison of various recording techniques, these short time intervals of between 30 and 120 seconds are effective. They are simple to calculate and therefore quickly generated. In the case of Figure 10.7, the activity would be analysed in both cases in the same way, if the foraging flight did not extend over such a long period.

However, these interval lengths fail when, for example, several individuals fly along a well-frequented flight path within a single time interval. In this instance, it is difficult to interpret the activity with an index, since the nature of the flight path is not picked up in this model. In general, the duration of the time interval determines the resolution of the index. Shorter consecutive activity phases of various individuals are counted as one activity.

10.4 Standardisation of the activity index

The activity indices described above are obtained by analysing the data generated by the acoustic recording. Individual datasets may differ, however, in the running time of the devices, because the base period of the data collection is not always identical. Data may have been collected over one or more nights. The devices may also have different running times. The duration of recording is likely to be shorter in June than in April or September, because the length of the nights varies with the time of year.

Therefore, the data have to be standardised when comparisons are made. The length of recording can serve as the base in this instance. Even more suitable, perhaps, is the length of the night, from sunset to sunrise. This standardisation to fixed time periods is relatively easy to carry out. If standardised figures are published, the standardisation must be highlighted and explained in detail. Only then will the reader be able to understand why some of the activity figures are very low. If the night is eight hours long, 80 passes will be reduced to 10 for each hour of night. If the reader is not aware of the standardisation, the activity will appear very low.

10.5 Quantitative analysis of activity

The quantitative analysis of activity is a constantly recurring problem in acoustic investigations. The question at issue is usually whether low, moderate or high activity levels were recorded. The indices described above (Section 10.2, *Quantification of activity*) are generally employed to answer this, in order to establish a figure which can be used for comparison. Such comparisons are used particularly in scientific research, but even in reports and assessments this kind of measure of activity makes an assessment so much easier. Having said that, the task of impact planning is not necessarily to compare locations, and, as a result, to allow development in the location with the lowest activity rate or number of passes. Rather the focus must be on the absolute risk and danger to the species that occur there.

10.5.1 Scientific investigation versus environmental impact assessment

In scientific investigations, the usage patterns of particular species in particular habitats are often studied and compared directly. For this purpose, it may be adequate to use a simple measure of activity for the description. If the issue is possible preference behaviour or niche behaviour by a species, the consideration of the activity times for each habitat is sufficient. However, a single evaluation may not adequately illustrate complex behaviour. It may not be enough to have a description based solely on the total level of activity derived from the time pattern statistics. A more sophisticated approach to this involves multimodal modelling, which requires very involved statistical calculations.

When carrying out research and investigation in the natural environment, steps must be taken to ensure that none of the species occurring in the study area is affected negatively in any way. Therefore, the analysis of the data obtained must be carried out as holistically as possible. Depending on the nature of the study, it may not be sufficient to use a simple index. The behaviour of the bats, for example, will cause the intensity of usage of the locations to change during the year. Proximity to particular roost types may result in activity peaks and troughs which are determined by season or time of day. It is therefore vital not to ignore the time element when evaluating the data in many of these types of study.

In comparison to scientific investigation, it is therefore important in landscape projects to include a description of the temporal components of bat activity and to take account of them in the evaluation. It is vital to take account of more than just simple habitat preferences in the analysis.

10.5.2 Consideration of time patterns

The standard types of index do not take account of the distribution of activity over time. An activity level of an average one recording per day can come about in many ways. Fifty out of 100 days with two instances of activity a day gives the same result as 100 days with one instance a day. The average is the same, so a comparison of the averages would result in both datasets being assessed as identical. As neither is an example of normal distribution, the usual statistical methods cannot be used to correct the analysis.

How then can the problem be solved without resorting to a time series analysis or multivariate statistics? One simple solution is a graphical representation of the time distribution accompanied by a written analysis. For example, when the activity of a species is being measured over several days, the concept of regularity or continuity can be brought to bear:

The species was encountered with great regularity. On the days when it was present, activity occurred on average about ten times.

A high degree of continuous presence implies that a species will be more severely affected by new development than one which uses that area only intermittently. It is necessary to be aware that even bats which are present on a fairly continuous basis may occasionally be absent for some of the time. This does not mean that the risk of disturbance is less. It is also important to bear in mind species-specific behavioural aspects. A good example of this is the need to take account of migratory activity in the planning of wind farms, especially when this is the only time of year with bat activity.

The distribution of activity in the course of the night may also be relevant to the assessment of a development project. If activity is always concentrated in one part of the night, this can have a bearing on the nature of the protection provided. Lighting in industrial estates, for example, can be switched off at the critical times such as when bats emerge from nearby roosts.

There is, however, no clear-cut threshold value between high and low levels of continuity. The implication of this is that a report cannot dismiss the need for measures simply because the level of activity is low at a certain point in time. Often the problem is that an assessment of zero or only low activity levels is a consequence of the extent of the investigation. If the location or area is not surveyed extensively, it is likely that the complete range of species will not be recorded. Even 15 or 20 walks around a site, which is the requirement in many wind farm guidelines, will not necessarily find the entire range of species present. Therefore, when stating that there is little or no activity, it may be necessary to acknowledge any potential shortcomings in the process (see also Section 10.1.1, *False negatives*, and Section 10.8, *The best activity index*).

10.6 Qualitative analysis of activity

Both in reports and in scientific publications, there is regularly an evaluation of recordings to determine whether bats are moving along flight lines, foraging or emitting social calls. The aim is to use qualitative analysis to supplement a simple quantitative index. This categorisation is based on the assumption that, by differentiating between call types, it is possible to highlight important locations and habitats because of the sorts of activity occurring there. Locations with recordings of numerous feeding buzzes, for example, are assessed as more valuable than those that have few or no foraging calls.

These assessments are only possible to a limited extent. Various factors affect the reliability of this kind of evaluation and hence the feasibility of making meaningful comparisons. Not all types of call are unambiguous and easy to recognise, as the volume of the types of call will vary, and the location of the microphone may play a significant role in determining whether some types of calls are recorded at all. Immediate identification in the field is almost impossible to achieve, and so recording will be essential. It is assumed in the following discussion that there will be a recording for each pass. Positive evaluations are possible, but the lack of evidence for particular behaviours is not proof that they do not occur at a particular location.

Even if the recording device is set up in a regularly used foraging area, it may be that on any one night the density of insects is low. Despite this, it is possible that bats will be detected regularly, but without any feeding buzzes. This demonstrates the difficulty of sampling a location on a single night.

A high level of continued activity without feeding buzzes does, however, strongly suggest that there must be foraging happening, albeit unsuccessfully. This location could then be justifiably categorised as a foraging area.

10.6.1 The range of call types

If different types of call can be distinguished and analysed, these must all be equally loud in order to make a comparison. The volume level is determined by the bat and is adapted to the purpose or objective of the particular activity at that moment.

Typical social calls during the mating season are usually very loud and of low frequency. They have to be effective over long distances and so can be detected more easily. On the other hand, there are other social calls which are intended for communication at close quarters and which do not carry far. A bat may also employ high frequencies, but at a low volume. Parent–offspring communication or contact calls, frequent in common pipistrelles, fall into this category.

The feeding buzz calls, which are used for locating prey, do not need an especially long range. They are used to precisely locate prey in the catching phase. As the bat will normally have already come close to its prey at this point, the range is relatively unimportant. The barbastelle, however, is known in effect to 'creep up' on its prey, so it is important to realise that the absence of feeding calls in recordings does not necessarily mean that the bats have not been foraging. Some species that pick their prey off vegetation, or find it by passive listening, generally do not emit feeding calls. Locations where these types of bats are feeding would thus not be recognised for their true value or purpose, if the occurrence of feeding calls was used as the sole criterion. Species that do not emit feeding buzzes include the *Plecotus* bats, Bechstein's, Geoffroy's and greater mouse-eared bats.

Not much is known about the calls used by bats when navigating along linking paths between roost sites or feeding areas; they are described only anecdotally. Since they are used like echolocation calls, the amplitude should be similar and will be equally easy to detect.

10.6.2 Recognition of various call types

Unusual or rarely encountered types of call must also be recognised, in order to allow analysis of recordings. This presupposes that there is a clear definition of call types, reflected in unambiguous measurement parameters. However, our knowledge of bat calls and the significance of each in terms of behaviour is still developing for the various species.

Although social calls can usually be easily recognised, it is not always clear when an echolocation call crosses over into a social call (Figure 10.8). The statistics for social calls are, therefore, a little imprecise. It is possible, indeed, to speak of cryptic social calls. A further problem is that numerous social calls have not yet been attributed to a particular species. While there is a good description of the repertoire for some species, there is a near complete absence for others. While qualitative descriptions may be undertaken, a comparison between locations is more problematic. For example, an area used for mating by noctules is easy to recognise, but a lack of clearly recognisable social calls in other locations does not equate to proof of a lesser or different significance for that site. Cryptic social calls, for example, could be overlooked. Similarly, higher-frequency contact calls used as mating songs may not be detected because of the lower distance range of these calls.

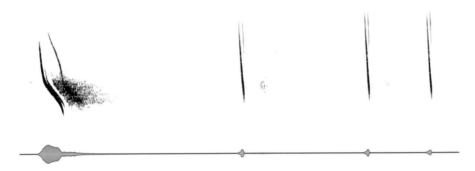

Figure 10.8   Social calls of Bechstein's bat followed by normal echolocation calls.

Feeding calls or feeding buzzes are generally simple to recognise. The bat will dramatically increase its repetition rate in order to precisely locate the prey. Equally, the calls will be much shorter in duration and will be strongly modulated. The frequencies drop as the gap between calls decreases (Figure 10.9). They indicate feeding activity, since they are used in the final approach to the prey. Sometimes, calls in the approach phase, preceding the moment of capture, are clearly recognisable, while the actual capture calls may be absent. This is generally interpreted as foraging activity. However, there are other situations in which a bat uses these quick bursts of calls in order to approach and avoid unexpected obstacles.

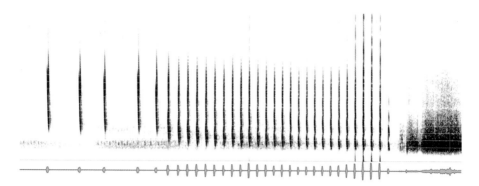

Figure 10.9   Call sequence of Alcathoe bat with feeding buzz and subsequent attempt to catch the microphone.

Calls used by bats on their way between roost and feeding sites are difficult to recognise, although it may be that these calls are of longer duration. With pipistrelle species, for example, these calls may be QCF (quasi-constant frequency) at the lower frequency spectrum for these species. However, this is not a generally recognised definition, as long calls are typical of open spaces. Even when a bat is flying along edge habitats, it may use this type of echolocation call if it turns to focus on the open space.

As well as the call shape and frequency, an increased inter-pulse interval is seen as an indicator that the bat is navigating between sites (commuting). Especially when the interval is about twice the duration of normal echolocation calls, it may well simply be the case that every second call has been missed. Calls will not be picked up because the bat has been directing calls away from the microphone (loud–soft rhythm, screening).

The recognition of the various types of call using automated techniques is not a reliable method. It is therefore necessary to use manual techniques, which will involve very lengthy checking, especially if a large number of recordings has been made. It also means that subjective errors will occur, compromising the results.

Mistakes will indeed be inevitable, and they will be difficult to spot or assess. This is especially the case if no particular interpretations have been made because of the absence of proof – for example, if feeding activity has been discounted because of the lack of feeding buzzes. Information on when species appear at the location can, however, provide useful information about the function of the site.

10.6.3 The effect of microphone location

The limited distance range of a bat detector means that the location of the microphone has a significant effect on the recordings made. For example, large accumulations of insects within the range of the device may lead to an increased proportion of capture calls in the recording. If the microphone is located in a different place within the habitat, away from the concentrations of insects, it can happen that no feeding calls will be recorded. The situation would be similar in mating areas, as an unfortunate placing of the microphone could result in all this activity being missed. The effect of the location on the data recorded has already been dealt with in Section 10.2.1, *The number of recordings*, and in Figure 10.7.

10.7 Comparison of data

It is particularly in scientific research that data are used to find differences based on the evidence of statistics. But evidence-based analysis can also be very helpful in reports or in the evaluation of habitat preferences. If recordings have been made in a consistent and standardised manner, it is possible to carry out comparative analyses of the results. Because of the difficulty in separating out individual bats (as already discussed), many established methods are not easily applicable. Nevertheless, some simple analyses can be carried out, which may allow better interpretation, even if not at a significant depth.

Moreover, based on the simple processes introduced below, it is also possible to apply complex statistical methods to research the data for provable differences. For the readers of this book, there will generally be no need for either component analysis or modelling, and so they are not dealt with further. Such matters are beyond the scope of this book.

10.7.1 Simple comparisons within one species

If the activities of a single species are to be studied in various locations, the activity indices described in Section 10.2, *Quantification of activity*, can be used directly. However, some errors may occur if the locations vary strongly in their structure. The bats adapt their amplitude or inter-pulse intervals according to how cluttered or open the location is. Often this is acceptable for comparisons. Potential sources of error in the indices have already been addressed in their descriptions and must be taken into account.

A standardisation of the data to take account of changes such as the length of the night or the duration of the recording is sensible if data are collected over a long period of time. Otherwise, a direct comparison of data in June (very short nights, many insects, warm) with September (long nights, fewer insects, cool) will provide an index with spurious differences, unless proper standardisation is carried out.

Thus, in September, many minutes of pipistrelle activity are recorded, because the bats need more time for foraging and the nights are also longer. They may then, for example, still be socially active for a certain time. In contrast, the density of insects in June is very high, and the same bat needs less time to feed and will rest or return to the roost, perhaps because it has to suckle its young. If the night duration is standardised, this distinction could again become much less clear (see Section 10.4, *Standardisation of the activity index*).

It clearly makes sense to assess the data not only according to a single index, but also to look at it from the point of view of activity during the course of the night (Figure 10.10). The interpretation will usually become easier and will better describe the actual activity recorded. By the use of these simple calculations, differences between locations can be discerned for individual species and portrayed in graphs. These are also very satisfactory for many aspects of impact studies, since there is no need for complex statistics.

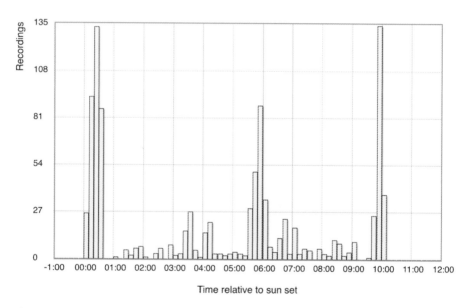

Figure 10.10   Example of a night-time activity pattern near a common pipistrelle roost. The graph is standardised to sunset at 00:00.

10.7.2 Habitat usage and species comparisons

The preference for a habitat or location by individual species can equally be established using relatively simple statistical procedures. The percentage usage of each of the locations or habitats studied is firstly determined by using the above-mentioned indices. The figures obtained for percentage usage can also be compared between species or used for other calculations.

Table 10.1 Example of a simple evaluation of the percentage usage of locations by three species of bats.

	Woodland	Woodland edge	Open country
Common pipistrelle	20%	50%	30%
Barbastelle bat	30%	60%	10%

For such comparisons, further calculations such as niche breadth, derived from the Shannon index, can be used. The basis of this calculation can be seen in Table 10.1, which shows the percentage usage for each location/ structure/habitat. With these percentage usage figures, the niche breadth (NB) is calculated for each species using the formula:

$$NB_j = \frac{1}{\Sigma_r p_{jr}^2}$$

where p_{jr}^2 is the percentage usage share of the species j in the resource r.

A standardisation for the comparison between species is derived from the following formula:

$$stand.NB_i = \frac{NB_i - 1}{r - 1}$$

where NB_i is the niche breadth for species i.

In Table 10.1, a standardised niche breadth of 0.82 is derived for the common pipistrelle. The values are 0.59 for the barbastelle bat and 0.43 for the greater mouse-eared bat. The mouse-eared bat has the lowest niche breadth in this modelled example.

A niche overlap between the species can be calculated in analogue:

$$NU_{jh} = 1 - \frac{1}{2} \sum_r |p_{jr} - p_{hr}|$$

In Table 10.2, the two species, *common pipistrelle/barbastelle bat* overlap by 80%. *Common pipistrelle/mouse-eared bat* overlap by 50% and *barbastelle bat/ greater mouse-eared bat* overlap by 60%. An application of this simple statistical device can be found in Volker Runkel's dissertation.

Table 10.2 Example of the niche overlapping developed with the formula above.

	Common pipistrelle	Barbastelle bat	Mouse-eared bat
Common pipistrelle	–	80%	50%
Barbastelle bat	80%	–	60%

10.8 The best activity index

The question is often asked as to the best way of assessing activity. The indices described in Section 10.2, *Quantification of activity*, all have their own shortcomings, and these are difficult or impossible to put right. That is particularly the case when, because of the device settings or the use of different devices, the recordings were made either without parameters or with different ones. If it is not recorded then it is not in the data. Nevertheless, even imperfect data can sometimes be used productively and should not be rejected out of hand. Depending on the nature of the study, some indices are robust, despite differences in the manner of recording. A rule of thumb is that the greater the impact of the technology on the data collection, the trickier it is to compare data which have been obtained in different ways.

If the indices that are strongly dependent on the recording technology (number of calls, recording time in seconds, number of recordings) are compared with a time-based index (minutes with activity, as discussed below), it will be seen that there is a correlation between all the indices (Figure 10.11). That is to be expected, since they all derive from the same activity. Not all parameters, however, correlate with one another in a linear fashion. While, for example, the number of recordings plotted against the number of seconds exhibits a smooth linear relationship, the number of minutes with activity plotted against either of the other two parameters does not. That means also that the activity does not have to take place clustered together, but that recordings occur again and again which have a gap of at least one minute between them. Only with large numbers of recordings does this effect disappear, and the initially steep curve flattens off (see last row of Figure 10.11). If the scatter of the individual parameters is examined more closely, clear differences can be seen (Figure 10.12). Since a considerable scatter makes interpretation of the data more difficult, indices should be used which have less scatter. From the graphs, it can be seen that data analysis based on a time interval (the minute classes in Figure 10.12) exhibits less scatter and is more robust for evaluation.

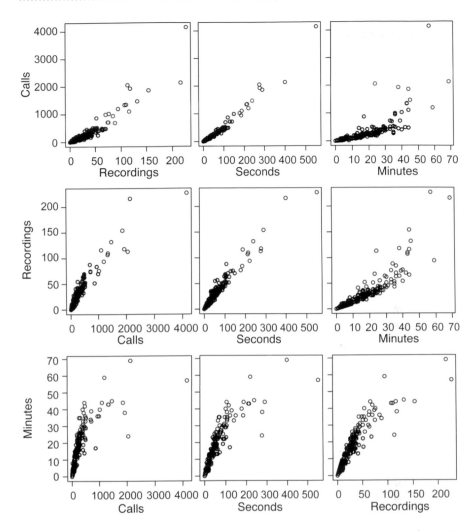

Figure 10.11   Based on 202 nights, the number of calls, recordings, minutes with activity and seconds of recording time are plotted against each other in pairs. The data were recorded using a batcorder with a threshold of −36 dB below the maximum amplitude and a post-trigger of 600 ms.

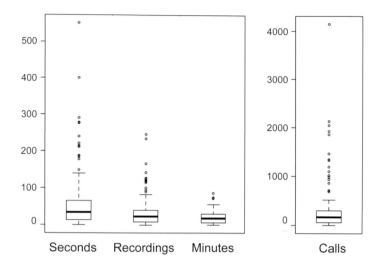

Figure 10.12 Boxplots with parameters of seconds, recordings, minutes and calls per night based on 202 survey nights (batcorder with −36 dB and post-trigger of 600 ms). Because of the significantly larger value range of the number of calls, this is represented separately on the right-hand side with its own scale.

As well as the technology used, parameters such as the exact location and the range of species also affect the indices. A species like the noctule emits significantly fewer calls per second than the common pipistrelle. The duration of stay measured in the recording area depends on the call amplitude and the style of foraging. Thus, the noctule will usually produce fewer recordings in a particular site than the pipistrelle, which hunts in confined spaces. This will result in different periods of stay (seconds) for each species. It is particularly the indices dependent on technology that are strongly influenced by such factors. Thus, an index that is relatively unaffected by technology and location is more suited to the description of activity.

The scatter or spread of the index is furthermore affected by the frequency or regularity of the data collection. Activity is more accurately captured if data are collected in a constant and continuous manner. While outliers will still be recorded, the average or median values will ideally be strengthened and come to the fore. For reasons of cost, data are often obtained by means of individual collections rather than long-term monitoring over several weeks. This raises the question of how much survey time is needed to obtain meaningful and useful data (see also Section 10.1.1, *False negatives*). From the dataset represented above (Figures 10.11 and 10.12), sets of ten recording days were selected ten times at random. In Figure 10.13, it can be seen that the scatter with the parameter *number of recordings*, with ten random samples (left half), is bigger than it is with the parameter *minutes with activity* (right half). An index with a small degree of scatter means that the results can be more effectively interpreted. Using an index based on the number of recordings, a very wide scatter will result when there is an extremely low number of recordings. Even the ten nights chosen in the example show considerable scatter for this parameter (see also 50% data ranges or the medians in the left-hand boxplots of Figure 10.13). Current practice in impact assessments does not specify 10 sessions, and evaluations are based largely on the number of recordings. This means that authoritative or meaningful results are not really possible. Simply changing to minute-based or similar time classes would result in an improvement.

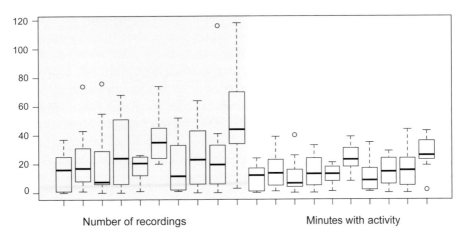

Number of recordings   Minutes with activity

Figure 10.13 From the dataset referred to in Figures 10.11 and 10.12, sets of ten recording days were selected ten times at random, and the distribution of the two parameters plotted, namely, number of recordings and minutes with activity.

As well as the simple indices, there are frequent attempts to evaluate activity by means of complex calculations. An index which is very laborious to calculate will perhaps produce slightly better results, but will mean a greater chance of error. Even with further processing of the data, such as a correlation with environmental parameters, simple processes with clear results are the methods of choice.

If data are to be evaluated comparatively by others at a later date, easy transferability is helpful. A complex formula may be misunderstood or incorrectly applied. On the other hand, there is no point using an index which is too simple and imprecise.

At all events, it is always important to define precisely what is being measured, irrespective of which index is used. This definition must be included in the report or research paper. It is only by doing this that other people will be able to interpret the data.

In order to undertake a simple comparability study and an evaluation which takes account of technology and behaviour, time classes with a length of a minute are recommended. The detailed comparison of different types of activity (foraging, social, passing along flight lines etc.) should be dispensed with. These factors can be drawn on in a qualitative description. If the contracting body stipulates the use of particular indices for analysis, then these must be respected. However, if their weaknesses prove to be a problem, a better index can be calculated to supplement the findings.

10.9 Dealing with large datasets in practice

The automatic recording of bat calls over a longer period of time increases the possibility of being able to carry out a precise interpretation. Distinctive phenological characteristics can be highlighted which would require an immense effort to replicate with manual methods. By means of direct sampling recordings, it is possible to analyse the range of species. However, large datasets can initially pose problems for the bat researcher. Many of these problems are discussed elsewhere in this book, and solutions proposed. Manual call analysis, in the case of large datasets, is particularly time consuming, and can limit the amount of automatic recording which is feasible during a season. Large datasets are thus best analysed automatically. While there are many advantages to this, there are also, of course, pitfalls. Above all, the results of automatic analysis must always be considered in the context of the recording location.

10.9.1 Identification of species and species lists

The initial step in the analysis of a large dataset is to draw up a species list. The manufacturers of automatic recording systems often develop special software which carries out the necessary tasks, particularly automatic call search, identification of species and presentation of the results.

In the second stage, after the automatic analysis, the user must check the results. This requires some experience with the software used. Normally there are settings to be chosen, which will in turn have effects on the results. The user must bear this in mind during the interpretation phase. Meaningful comparisons with other surveys, if this is the intention, are only feasible if the hardware and software settings are the same.

With practice, the user will gain an awareness of the potential misinterpretations that the software is prone to. Equally, it will become evident which species are usually correctly identified. That is very helpful for a rapid first viewing of the dataset. It will then be possible to make a quick interim assessment. If it transpires that there are species that are difficult to tell apart, then the decision can be made in good time to capture some using nets or harp traps, if precise species identification is needed for the study.

It will also be necessary to take a sample of some of the trickier species to analyse manually. Even at this early stage of the study, all the calls, especially those of the critical species, should be checked manually. In this context, it is important to assess how probable it is that the species occurs in the area. Obvious misidentifications by the software do not need to be checked until later. This should be done at some point, however, as it may be a species that was not expected in the area, but which is expanding its range.

The most important factor in this second step of the analysis is, in fact, the human operator. Their experience, and not only with the software, has the greatest effect on the interpretation of the results. Although, at this early stage of the analysis, the time and effort needed are still small, there is always a need to look critically at automatically generated results.

As well as a simple list of species, the creation of a species tree is always of practical value. It is a way of looking more closely not only at the frequency of passes or the minutes with activity, but also at the recordings which have not yet been narrowed down to the species level. If there is a reliable breakdown of an acoustic group into a range of species, the recordings which have not yet been narrowed down to the individual species level will very likely belong to species that have already been identified automatically.

Important findings, which are liable to substantially influence the progress of a project, can often be achieved by the simplest of means. The routine after a change of the memory card should therefore include not only sorting and backing up the data, but also rapid, automatic analysis and interpretation and a first processing by an expert. This is quite feasible in periods when field work is the main activity.

10.9.2 Species phenology

After the thorough analysis of a dataset, often at the end of a season, there will at least be a table of results available which is now independent of the software or method used to generate it, and which can be uploaded into a spreadsheet for further processing. There is now the opportunity to represent the data in graph form, or in diagrams which relate to specific aspects such as the phenology of the bats. However, this process often involves considerable time and effort, since spreadsheets are not designed for the specific require-ments of wildlife investigations. In this context, the manufacturer's software is often of great help. Since it is common to be processing large quantities of data, it is worth using statistics programs to exploit and present the data.

One of the great advantages of long-term recording is the opportunity to describe the phenology of both local and migrant species. The relationship to sunset and sunrise together with the occurrence of the species over the course of the year can be seen at a glance. A kernel density plot allows the depiction of phenology even if the passes took place less often, since it portrays the relationship of the activity to particular periods of time.

An example of this can be seen in Figure 10.14. Here the individual recordings are plotted both as points and as kernel density clouds. In summer, the passes noticeably occur after sunset and before sunrise, indicating that there is probably a roost nearby. There is a further concentration visible in autumn, which suggests that there is a mating roost in the vicinity, and this in turn may be evidence of reproduction and migration.

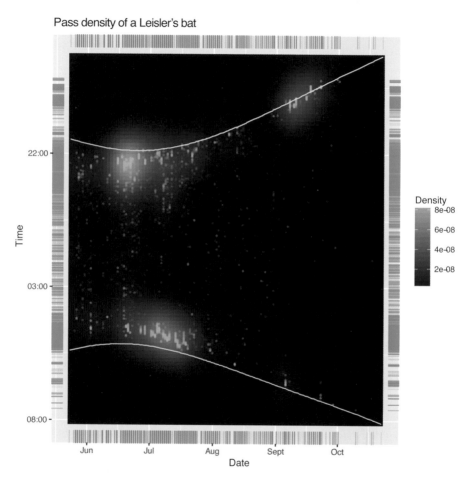

Figure 10.14   Possible representation of the phenology of Leisler's bat from a long-term monitoring project.

10.9.3 Why do bats use particular areas?

Apart from carrying out counts in the field using visual observations, automatic recordings are a useful tool in investigating for what purposes a space might be used by bats. The previous example of phenology (Figure 10.14) illustrates the use of the area for roosting and reproduction.

For example, if there is a large number of structures in any area that are part of a flight line, then automatic recording systems can be installed as a supplement to the visual observations where observation might otherwise be difficult. The interpretation of the data will make it easier to decide where visual observation should occur in future surveys. Analysis of the data will help to highlight important aspects of how bats use an area. The following are the main behaviours, together with short explanations of how to spot them in the data:

- **Roost** – when, depending on the species, increased activity is detected close to sunset or sunrise (Figure 10.15).
- **Reproduction** – when, in a known mating period, an increasing number of appropriate social calls are recorded.
- **Flight route** – where there is considerable activity shortly after sunset.
- **Foraging habitat** – where, depending on the species, continuous activity, particularly feeding buzz sequences, are recorded throughout the night, or at particular times.

When the presence of a roost is suspected, it is particularly important to verify and precisely locate it using other methods.

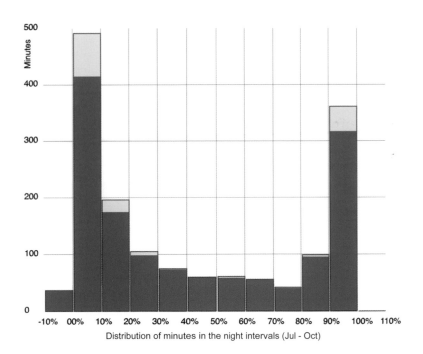

Figure 10.15 Depiction at intervals during the night of the activity of a Leisler's bat near its roost. The night was divided into 10% intervals and each recording assigned accordingly. The 10% classes were adopted to avoid distortion resulting from the changes in the length of the night as the season progresses.

11 Quality assurance of reports

In impact assessment, the specialist report serves to determine to what extent bats will be affected by a development and how its impact can be mitigated. The need for impact assessments is usually enshrined in national or European law. The specialist report has great significance, as its purpose is to ensure the protection of the bats. Great care therefore needs to be taken in its preparation. Unfortunately, in practice the necessary quality is often lacking. One of the reasons for this is that there is a dearth of specialist training and an absence of clear, independent guidelines. Even when guidelines are provided (in the UK these include the Bat Conservation Trust's *Bat Surveys for Professional Ecologists: Good Practice Guidelines*), reports are often massively inadequate (Gebhard *et al.* 2016; BUND *et al.* 2017).

Personal experience and subjective judgement are key factors in the analysis. Authors of reports do not always follow the procedures. This may be due to a lack of knowledge on the part of the author and a failure to check facts and methods. This makes independent, neutral checking more difficult. It is therefore important to adhere to certain rules during the production and writing of a report, in order to increase its quality, and to make easier any testing by authorities and courts.

In this chapter, criteria are presented which can be used to improve the quality of specialist reports derived from acoustic data. The chapter provides details of information that should be included, and procedures to follow, in order to ensure a professional and accurate report. If certain information is missing, interpretation becomes difficult or impossible. Furthermore, it could result in legal problems in the event of a lawsuit or review. The suggestions have been framed so that the report documents clearly how the data were collected. If questions arise over the collection of the data, it will be possible to show clearly both how the data were obtained and which data have fed into the analysis.

This chapter makes proposals for the structuring and presentation of the report, as well as its verification. However, it cannot cover all eventualities, and there may be special additional requirements which necessitate a deviation from the standard content and structure.

11.1 Devices used and their characteristics

When writing a report, it is essential to provide accurate details of the devices used and their mode of operation. Only then can the reader judge whether the data were collected in the right manner, and whether there might potentially be errors.

11.1.1 Types of device and versions

Every device which is used must be documented precisely. Details of the bat detector need to be given, including the manufacturer, the exact name and model of the device and, if relevant, the version of the firmware. This will enable readers and users of the report to understand the characteristics of the device used, and to draw inferences about the data collection. For a direct sampling detector, the information might look something like this:

ecoObs batcorder 3.1, Firmware 307

An example of a time expansion detector:

Pettersson D240x

If the device has various add-ons or microphones (e.g. SM2Bat, Batlogger), these should also be included:

Elekon Batlogger, FG-Black Mikrofon, Firmware 1.2.3

If several devices of the same type are used, each device should be clearly documented (inventory number, serial number etc.), so that, in cases of doubt, it is possible to trace which device was used for a particular survey.

11.1.2 Settings

As a rule, devices allow a wide range of possible settings. These must be documented with each recording. If the setting is never changed during a survey, then it is sufficient to mention it only once. Every different setting needs to be described so that users who are unfamiliar with the technology used can understand the purpose and effect of each setting. Often it will be possible to use the text from the instruction manual provided with the detector.

However, in the case of simple devices, such as heterodyne detectors, this documentation is difficult. It is not really possible to document any settings, especially when a device is used manually for transects, with the settings continually being changed. However, some relevant operational information should be provided:

- Use with/without headset
- Rate of change and frequency range, for example, within about 10 seconds between 18 and 55 kHz

If devices have a control dial, with or without latching, the exact setting of the dial should be noted. If values can be selected in menus, these should be listed in precise detail. The failure to provide such information is negligent and prevents interpretation of the data collected. In the case of monitoring on wind-turbine nacelles, the lack of this information could have very negative consequences in a court case, possible resulting in a requirement to repeat the monitoring.

11.1.3 Status of the microphone

Apart from the device settings, the status of the microphone is particularly relevant to the recording, and this information needs to be provided. This is not straightforward, as there is effectively no reliable gauge for this. If a calibration has been carried out, the date and name of the person who did it can be given. If there has been no official calibration, for example by the manufacturer, then providing its age will give an indication as to the sensitivity of the microphone. If the microphone has been tested, this should, of course, be logged, together with precise details of the nature of the test. Jangling keys, by way of testing the microphone, may be sufficient as a general test, but it does not allow precise conclusions to be drawn about the microphone.

11.2 Recording methods

The results of the activity recording depend not only on the devices used, but also on the way those devices are employed. This, too, needs to be documented.

11.2.1 Recording location and setup

If data are obtained using passive detection systems, then the location and setup of the device will have an impact on the results. As precise a description as possible is necessary for the correct interpretation of the results. For example, woodland species will not be expected in open areas. Conversely, species that use open country will only occasionally turn up in woodland with a closed canopy. The description of the location should therefore contain information on the landscape structure. A single sentence will usually suffice. In addition, the setup of the device should be explained and illustrated. The various methods employed to protect the device from the weather will also have an impact on the results. A device concealed in a nest box will produce different results from one located in the open.

With recordings made by walking along a transect, GPX tracks should be noted instead of location descriptions and they should be shown on maps in the report. Details of the exact methodology used should also be given (transect protocol).

11.2.2 Recording period

This includes, among other things, precise details on the dates, duration and frequency of the recording periods. Any interruptions or discontinuations in the survey walks must also be documented. In the case of long-term monitoring, any downtimes must also be clearly indicated.

The information must be laid out so that it can be easily understood by the reader. It is not important whether relative times (e.g. *recording always for three hours from sunset*) or absolute times (*recording in July starting at 21.00 and ending at 6.00*) are provided. With relative times, it is only important that the reader can easily convert them into absolute times.

11.2.3 Climatic conditions

Details of the weather conditions are potentially of great value, as that will allow subsequent assessments to be made of the data. A very simple and straightforward way of expressing this is:

> The recordings only took place in weather conditions when bats were likely to be present.

However, this is plainly inadequate. The statement is too imprecise and allows no interpretation of the actual recording conditions. An improved version would be:

> Recording took place on nights when there was no rain, little or no wind and the temperature exceeded 8 °C.

While this is more precise than the first simple version, there is still scant information for the reader. Nobody would normally carry out recordings in pouring rain, gusty wind conditions or frost. It is therefore more meaningful to provide the minimum and maximum temperatures together with wind and rain thresholds. For long-term recordings, it is essential to give the temperatures over the entire period of the survey. All this information is not only needed for the interpretation of the data by the report writer, but also allows readers to make their own judgements. Temperature can be measured using temperature loggers or recording devices which include this information directly in the recordings. Good software will usually allow an analysis and presentation of this kind of information in the form of graphs or tables.

11.3 Analysis of recordings

The evaluation of the recordings is an important step in the production of a report. It is at this stage that the recordings are attributed to a species or genus. The assessment of bat activity and the impact of the development is based on these results. It is therefore essential for the report to make it clear how the recordings have been evaluated, whether it was entirely an automatic analysis, or whether samples or all recordings were manually identified. In the latter case, the process and the choice of samples must be clearly defined.

With manual checking, the recordings are usually presented in the form of sonograms. Here, information must be provided on the software and the version used. The settings used for the sonogram must be documented, as these have an impact on the species identification process. Reference works are often used in the checking process, and these must be explicitly named. Any rare species should be included in the report in the form of sonograms, either of single calls or of sequences of calls.

11.4 Information on activity

As has already been discussed in Chapter 10, there are many ways of analysing the data obtained and describing it in the form of an activity index. It should always be made clear which index is being used and how it is constructed. This ensures that the reader can understand what the data are actually showing.

11.4.1 Effects of the technology used

The technology used will usually have an effect both on the data and on the analysis. It should never be assumed that the reader of the report will be acquainted with the technology, and so it is vital to write a short paragraph summarising how the device works.

If the number of recordings is given as an index, it must be clear how the device generates recordings. If, for example, it always records for 60 seconds after being triggered, the reader needs to be aware of this in order to be able to interpret the data.

If the report does not show how the index is to be interpreted in relation to the recording device used, readers will not be able to relate the data to existing empirical evidence. An impact assessment would not then be possible, based on the data provided. This would represent a serious short-coming, and in the worst case the conclusions drawn from the data would be declared invalid, and new analysis, or even new recordings, would become necessary.

11.4.2 Calculation of an index

If a complex index has been created, it is a good idea to incorporate a small example of its calculation in the report, using a partial dataset. This will allow the reader to understand how the results have been produced. It will also make it easier to reuse a useful index in other reports.

If the index has already been published, and the definition is freely available, then it will be enough simply to refer to the source without a lengthy explanation. It is vital to ensure, however, that the reference source is still current and accessible. Otherwise, the lack of transparency will mean analysis is not possible.

11.4.3 Presentation of the results

The results should be presented in an easily digestible form, and there are certain principles which should be respected. On the whole, the same general framework can be used as for a scientific publication.

A simple graph showing the activity over the course of the year (Figure 11.1) or the course of the night is very effective as an overview. This type of graph is often a standard part of some analysis programs, and thus simple to produce. This is by far the most straightforward way of visually presenting the data obtained from long-term studies such as wind-turbine monitoring.

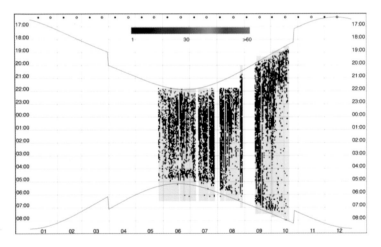

Figure 11.1 Simple representation of data from long-term monitoring over the course of the year. The graph shows the points in time of the recordings, expressed in 5-minute intervals between sunset and sunrise. This highlights the times of the night and the year at which bat activity occurs. The grey shading shows the operating time of the detector.

A graph of this kind may not allow the presentation of detailed activity statistics, but it does permit the reader to spot activity patterns very quickly. Alternatively, a bar chart can be used to illustrate the activity over the year (Figure 11.2). The numerical differences in activity can be easily seen with this method, even if part of the time component during the night is lost. It is important to highlight any gaps in recording.

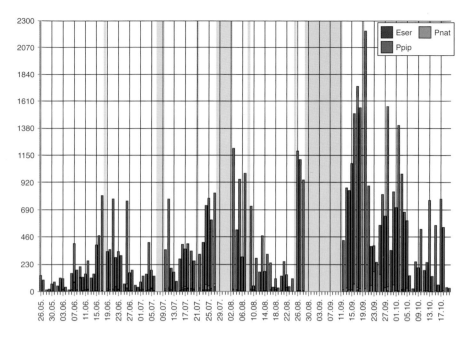

Figure 11.2 Bar chart showing the activity over the course of the year of serotine bat (Eser), Nathusius' pipistrelle (Pnat) and common pipistrelle (Ppip). Each bar represents the total activity of each species for each day. Gaps in the data are highlighted by means of a background colour, in order to distinguish between nights without recording from nights without activity.

Any tables showing all the recordings or passes are best placed in the appendices. Moreover, tables like these are almost never simple to read and do not provide a good overview of the data. For that reason, they could be dispensed with, or added later if the need arises.

Other diagrams should use data related to particular species or genera. These should be selected to answer or illustrate key questions in the investigation.

Pie charts which directly compare the activity of different species are best avoided. Direct comparisons of species are only valid in exceptional cases when the species call loudly and pursue similar foraging strategies. The unevenness of recording means that perceived differences are often simply methodological artefacts. Thus, the validity of such comparisons is extremely limited.

12 Nacelle monitoring – benefits and limitations

Although the acoustic monitoring of the nacelles or engine pods of wind turbines has become established as a well-proven method, there is an almost complete lack of discussion on the limitations of the technique. There is a need to look critically at many aspects of its application in practice. The recording range and thus the area recorded is limited for technical and physical reasons. Because of the great length of modern rotor blades, a single microphone on the nacelle will not necessarily give sufficient coverage. Moreover, measuring acoustic activity by means of short recordings is an issue that needs further consideration.

This chapter seeks to throw light on some of these aspects as they are directly or indirectly linked to acoustic recording. However, lack of space means that there are other contentious issues that are beyond the scope of a book on acoustics.

The comments refer to the dataset and analysis of the Renebat projects (Brinkmann *et al.* 2011; Behr *et al.* 2016, 2018), which are regarded as a benchmark. Measures for the reduction of bat mortality were established and a software tool (ProBat) developed from these projects.

12.1 Recording numbers and duration

The analysis of activity in the Renebat project was based on the number of recordings. As described in the Section 10.2.1, *The number of recordings*, this measure of activity is not particularly suitable for comparisons as it is heavily dependent on the technology used as well as on the behaviour of the bats. If the device settings are programmed incorrectly, the analysis of the results can be unreliable or wrong. In particular, there is a risk that incorrect settings will lead to an underestimation of bat numbers.

The project recordings were primarily carried out using the batcorder, with short post-trigger settings of 200 ms. Call sequences of bats with long inter-pulse intervals are thus recorded in many small files. The analysis of mortality risk, both in this project and by ProBat, is based on the number of recordings, so that, in simple terms, a high number of recordings will suggest a high mortality risk. Conversely, a low number of recordings will imply a low mortality risk.

The boxplot in Figure 12.1 shows data from acoustic recording at a nacelle. The overwhelming majority of the nyctaloid recordings (*Nyctalus*, *Eptesicus* and *Vespertilio*) have a recording duration of 200 ms. Although longer recordings did occur, they were rare and can be seen in the boxplot as outliers. The pipistrelloid species behave differently, emitting calls with significantly shorter gaps between calls. There are outliers here too, but there is a clearly distinguishable group with over 50% of the recordings with a duration of over 200 ms.

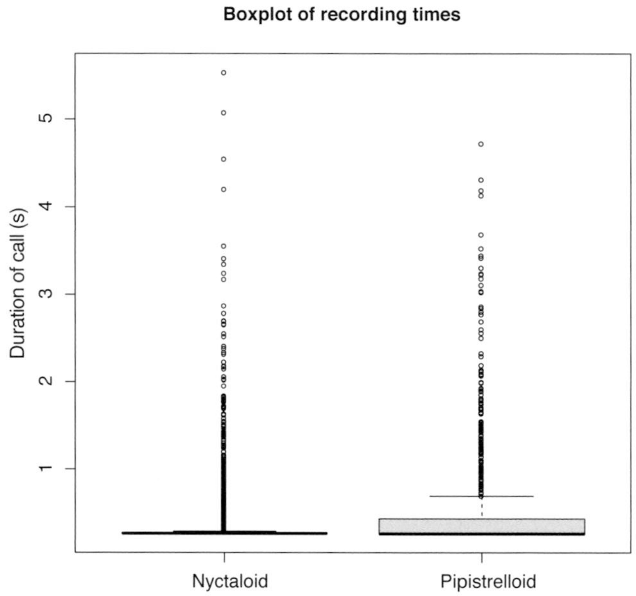

Figure 12.1   Call durations of nyctaloid and pipistrelloid bats recorded at a wind-turbine nacelle during the Renebat project. Since *pipistrelloid* species produce calls with shorter intervals between calls than *nyctaloid* species, they will have fewer but longer recordings.

It is entirely justified to consider both groups jointly in the analysis of the effects of operating wind turbines. It is insignificant which species is present at any one moment in the danger area of the rotor blades. However, if the activity is analysed in terms of the number of recordings, the assessment of the danger will vary depending on whether there is a high or low level of activity of pipistrelle species at the site. For any one period, the increased presence of pipistrelle species will mean that fewer recorded files will be stored than when nyctaloid species predominate. As the ProBat dataset differs from the Renebat dataset in its species mix, it has to be expected that there will be a different rate of error and that the ProBat dataset may therefore lead to an underestimation of the mortality risk.

In order to reduce scatter or spread of the data, which in the assessment of the mortality rate is responsible for the lack of accuracy, it would be better to use another measure of activity. A consideration of activity within short time classes of, say, between 30 and 60 seconds would have the positive side effect that the activity would tend to portray individual bats. This would make it possible to reduce dependence on particular recording systems. The usage of intervals also allows the inclusion of data sampled at locations like the lower blade tip.

12.2 The effect of rotor diameter

The wind turbines that were investigated by the Renebat Project, and which now serve as a benchmark for mitigation measures, mostly had a rotor diameter of 70 m and, on average, a hub height of 98 m. Since this research project, the typical wind turbine configurations have radically changed. Above all, this relates to the rotor diameters, which have increased in size to 141 m (Enercon) or to 131 m (Nordex). This corresponds to a change in the swept area from 3,848 m² to 15,614 m², an increase of 400%. The range of the detectors cannot necessarily keep up with this, since there are physical and technical limitations (see also Section 9.3, *Recording distance and amplitude*). At best, the range is a maximum of 50–60 m for the low-frequency calls of the noctule and 25–35 m for the common pipistrelle. Usually, the actual ranges are much lower. This depends on factors such as the amplitude and direction of the bat calls. Because of the uneven shape of the sound lobe, the direction of the call also has a significant effect on the likelihood of a recording being triggered. If the bat does not call directly into the microphone, the recording range will decrease significantly (see Figure 8.2).

While the required detection range can be more or less covered in the case of a 70 m diameter rotor, a not insignificant proportion of the swept area will be missing with the large-diameter rotors. Figure 12.2 shows this in a simplified form for the common and Nathusius' pipistrelles (see also Figure 9.4 for the sound lobe of the microphone). The diagram shows both the sound lobes of the bat echolocation calls and the acoustic shadow effect produced by the nacelle itself. Despite the large area of acoustic shadow, the results of Renebat II (Behr *et al.* 2016) seem to indicate that, with rotors of up to 80 m diameter, the recording is sufficient to predict the mortality rates.

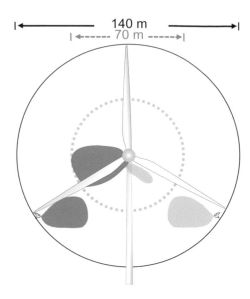

Figure 12.2 The diagram shows the recording area of a microphone installed on the nacelle for a common or Nathusius' pipistrelle. The sound lobe of the bat is shown both for a bat approaching from the left (microphone side) and for one approaching from the right. The recording of the bat flying from the right is significantly impaired because of the sound shadow cast by the nacelle.

12.3 The effect of hub height

There are two different trends with regard to the heights of towers. Towers with hub heights of 140 m or more are now being constructed. The rotor-free area of these towers, namely the distance between the ground and the lower tip of the rotor, is anything from 50 to 80 m, and does correspond to the benchmark. However, because of planning aspects (spatial organisation, height limitations), large rotors are also being built on shorter towers, particularly in areas of low wind. Compromises are often made, with large rotor diameters to increase the output and low hub heights to satisfy the requirement for a stipulated distance from human habitation. One example of the new configurations is the Nordex N117 with a hub height of 91 m and a rotor-free area of around 32 m. Even lower models are also available, with the tips of the rotor blades only 20 m above the ground. These kinds of turbines have been in regular production since at least 2017.

This means that the swept area will include zones where there are species that are normally categorised as slightly prone or not prone to collision risk (see also Figure 12.3). These will presumably also call more quietly than at altitudes of 60 or 100 m where landscape structures are usually out of range of the echolocation calls. The low-flying bats will mainly direct echolocation calls downwards at interesting landscape structures where prey is most likely to be found. It is the quiet calls of the pipistrelle species that will be overlooked or insufficiently monitored, particularly the common and Nathusius' pipistrelles, which are often the victims of collisions with wind turbines.

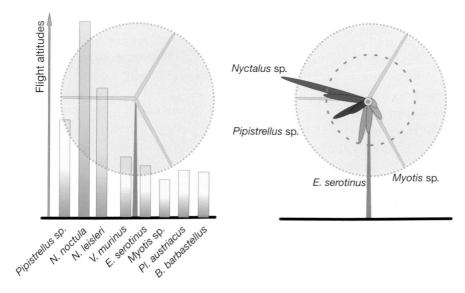

Figure 12.3 If the height of the rotor above the ground is low, there is a potential danger to species that otherwise rarely or never fly into the rotor-blade area. With nacelle monitoring, these bats cannot be reliably recorded because they normally produce quiet calls, and so cannot be easily detected. Shown here is a Nordex N117, and, on the right-hand diagram, an E70 shown as an inner dotted circle.

These factors will all reduce the likelihood of detecting these species, since the detector is over 50 m away in the nacelle. A reliable detection of bats in the area where the tips of the rotors are at their lowest is therefore unlikely, and so, unlike in the Renebat project, nacelle monitoring alone will no longer ensure the effective measurement of activity.

12.4 Improving the methodology

One possibility for compensating for the recording blind spots of nacelle monitoring is to install a second microphone to monitor the area of the lower rotor tip. This can be attached between 10 and 20 m below the lower tip of the rotors in order to avoid damage to the microphone and reduce any interference from the rapidly moving rotors (potentially several hundred kilometres per hour). If the rotors are very low, it is a good idea to reduce the recording sensitivity, so that the number of bats recorded hunting low above the ground and out of reach of the rotors is kept to a minimum. Alternatively, the detector can be operated at high sensitivity with lower-amplitude calls being filtered out at the analysis stage. Care must be taken to avoid deflection of sound upwards from reflecting surfaces, which will lead to double recording (see Section 9.6.1, *Sound deflection*). As there is still a lack of benchmark data, it is unclear how exactly the data from a second microphone should be incorporated into the analysis.

Ideally, it would be best to discount all the recordings of the second, lower microphone which are also recorded at the nacelle within a short time window (e.g. one minute). A comparison of the two datasets can subsequently be carried out in order to show the relationship between activity registered in the upper and lower recordings. This will allow a direct assessment of the blind spots of the nacelle monitoring.

Furthermore, the influence of wind speed and temperature on the distribution of activity in the lower part of the swept area can be determined. These two values must always be related to the nacelle values of the SCADA system (supervisory control and data acquisition), as it is only by using these that control of the turbines will be possible. Figure 12.4 shows analyses which may be carried out with these datasets.

12.4.1 Acoustic deterrence

Studies carried out at road construction sites have shown that bats avoid noise. If the noise drowns out the echolocation calls, then bats may go as far as to avoid the area altogether. Scientists in the USA have been trying to use this principle near wind turbines (Arnett *et al.* 2013). Loudspeakers fill the area around the rotors with noise so that bats will no longer fly into the danger zone. The idea is that the rotors will then not need to be switched off to reduce bat collisions.

The underlying idea is to mask the echolocation calls with random ultrasound noise and disorient the bats. Areas that are harmful to bats can be made unattractive to them. The initial results from these US studies show that the collision rate can be reduced at turbines with small rotors.

It would be useful to carry out a similar study in Europe. While the disturbance principle will almost certainly apply to all species, hearing thresholds and tolerance of disturbance will clearly differ between species. The authors of the American study detected a variable effect, particularly with respect to the frequencies used by bats. Furthermore, the possibility cannot be excluded that bats will get used to the disturbance and adapt to it.

For acoustic deterrence to be effective, the sound signal emitted must be loud enough to disturb the echolocation of the bat and cause it to avoid the area. Assuming a rotor length of 70 m, the acoustic deterrence signal must travel 70 – 90 m, and still be loud enough at that distance to remain effective. The authors of the US study see this as a problem, as they estimate that, in

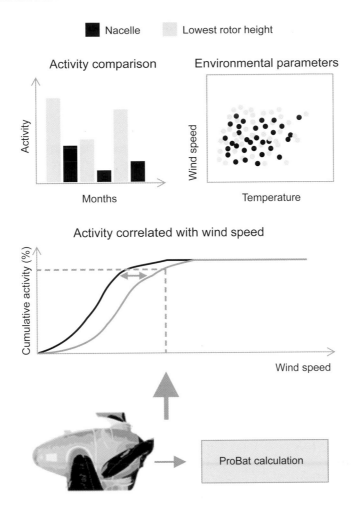

Figure 12.4 Some possible approaches to the analysis of data from recordings both at the nacelle and at the lower part of the swept area of the rotor. Important here are also direct comparisons of activity with respect to the height at which activity takes place as well as activity correlated with weather variables. Based on the differences established, the results of a ProBat calculation could be adapted manually.

order to disturb its echolocation, at least 65 dB will be needed at the height where the bat is flying.

Once the costs of the technology, the supervision of the equipment and the power consumption of the ultrasound loudspeakers are taken into account, it can be seen that this is not a feasible deterrence measure for modern wind turbines. The loudspeakers will have to be continually checked to ensure they are working. To do a daily check of each loudspeaker will require a sophisticated system of supervision. As the rotors become bigger, the sound will need to be played at a higher volume and the number of loudspeakers increased. A decision will have to be made as to where the loudspeakers should be located to give a good coverage of the rotor area. There may be practical limitations as to where a device can be placed on the nacelle.

All this will involve not only increased equipment costs, but also higher electricity consumption. It is important to check whether the overall balance of costs is more favourable than a curtailment algorithm which can trigger a real-time shutdown of the operation of the rotor blades (see Section 12.4.2, *Shutdown in real time*). In the final analysis, it has to be ascertained whether installation of a device and the supply of electricity is a viable and reasonable solution. The nacelle will have to be opened up and the devices will have to be adapted according to which of the many types of nacelle is installed. The nature of the different wind turbine sites also varies greatly.

Because of the physics of sound propagation, deterrence is not always possible with bats that use higher frequencies, above all when the necessary distance range increases. This is, for example, only possible with a nacelle-based system if the output level is well over 200 dB SPL (measured at 10 cm). So far this has not been technically possible.

Currently, experiments are being carried out with loudspeakers installed on the rotor blades as an alternative to fitting them on the nacelle. The advantage of this is that the range of the loudspeakers and thus the volume of the signal do not need to be as high. However, a correction will have to be made for the Doppler shift according to the position of the speakers on the rotor blade and the rotational speed of the rotors. The deviation caused by the Doppler shift depends on frequency and can amount to as much as 10–15 kHz. In order to eliminate the acoustic shadows created by the rotor blades, the loudspeakers will have to be installed at the very least on both the front and rear sides.

If a sufficiently large acoustic deterrence scheme can 'scare off' the bats from the area around the rotor, this area will be lost to bats. The operator of the wind farm would have to provide alternative areas as compensation. In contrast, this would not be necessary if a curtailment algorithm was used by the wind farm, as the area could still be exploited by the bats.

12.4.2 Shutdown in real time

Real-time shutdown of the rotors when bat activity is detected by means of acoustic methods is not wholly effective when the rotors are very big, for the very same reasons that acoustic recording is unreliable under similar conditions. The failure to record bats near the tips of the rotor blades means that there is no effective protection of some bats. Again and again, there are casualties which could be avoided.

Even if the detection range on larger rotors could be extended sufficiently, the time taken to shut down the turbine is too long. Bats would still be flying into the danger zone, before the rotor is turning slowly enough (Figure 12.5).

Detection would have to occur well before the bat enters the rotor zone. With a detection device in the nacelle this is almost impossible. Even microphones in the blades would not supply reliable data because of the potentially high rotation speed.

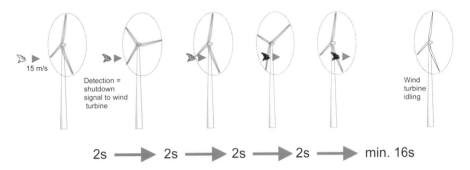

Figure 12.5   The process of a real-time shutdown and the flight path of a bat during this period of time. Before the wind turbine is at idling speed, the bat has long since flown through the still rapidly turning rotor blades, and is thus at risk of a collision.

With direct installation in the nacelle, static noise can lead to false stops. The operation of a wind turbine produces many noises with levels of up to 106 dB, some of which are in the ultrasound range. Some noises sound like bats and thus lead to false-positive identifications and result in unnecessary shutdowns.

One possible remedy would be a microphone ring around the periphery. However, this would quickly become very costly because modern turbine configurations require acoustic surveillance on at least two to three levels. The costs for the installation of masts and detectors for an operational life of up to 20 years should not be underestimated. Furthermore, technical aspects such as the networking and integration into the SCADA system would also have to be implemented.

13 Bat calls

Bats use ultrasound calls to get their bearings, to navigate and to search for prey, using so-called echolocation. The calls of bats differ from each other with respect to parameters such as frequency, length and shape of call, and inter-pulse interval. The call is designed to optimise the bat's perception of its environment and its prey. And because many different bats have similar requirements, the calls of different species and genera of bats are sometimes difficult or impossible to distinguish at the analysis stage.

As well as using echolocation calls, bats also communicate acoustically. They use so-called social calls to draw attention to themselves in courtship or to exchange information in roosts. There is mostly, although not always, a clear separation between social and echolocation calls, the former being longer and lower in frequency.

This chapter provides a short overview of bat calls, but it can neither provide a detailed description of the science of echolocation nor be an identification guide. Its sole purpose is to give a brief insight for the interested reader.

13.1 Call types and their function

As shown in Figure 5.2, bat calls are broken down into CF, QCF, FM-QCF and FM calls. This distinction is made on the basis of the shape of the call. Calls can contain more or less just one frequency (CF and QCF), or exhibit a continual change in frequency (FM), or be a mixture of both types (FM-QCF). While there is no universal definition of the individual call types, the following are useful guidelines (see also Figure 13.1):

- CF call: The average rise in the call frequency is <0.1 kHz per millisecond.
- QCF call: The average rise in the frequency of the whole call is ≥0.1 kHz and <1 kHz per millisecond.
- FM-QCF call: the call contains FM and QCF sections, each of which are at least a millisecond in duration.
- FM call: the call contains no sections whose increase is <1 kHz/ms and which are longer than a millisecond.

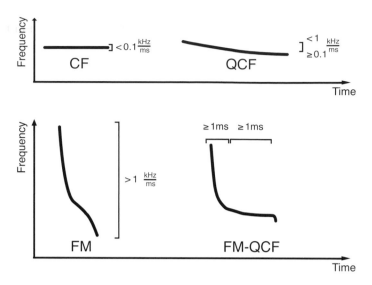

Figure 13.1   Four types of call can be distinguished on the basis of the average steepness of the call shape.

Figure 13.2 illustrates the different types of calls with examples of European species. It must be emphasised that some species are capable of various types of call and so do not follow this pattern. The genera *Nyctalus*, *Eptesicus*, *Vespertilio* and *Pipistrellus*, in particular, are able to emit CF, FM-QCF and FM calls. These calls are used flexibly according to the needs of the bat in any particular situation. This makes it difficult, or sometimes impossible, to identify the species of bat.

Calls with a wide bandwidth, FM calls, improve the ability of the bat to measure distances. This is particularly important in confined areas. In contrast, QCF/CF calls are more effective over longer distances because the energy in a single frequency is greater. This makes them suitable for foraging in open environments.

The following sonograms illustrate the different call types and their flexible use by individual bat species. The frequency range shown is from 0 to 150 kHz. Higher frequencies are not included. Call pauses are cut out of the sequences to provide a clearer picture.

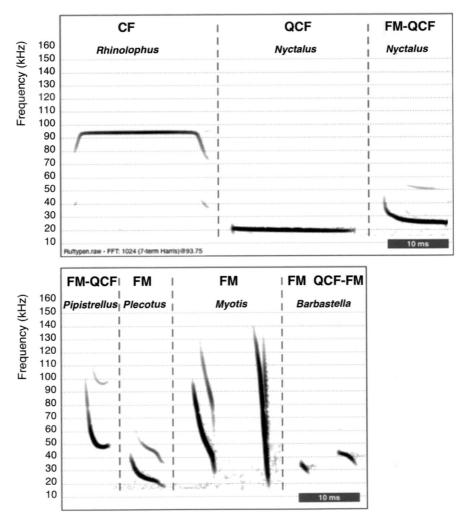

Figure 13.2 Examples of echolocation calls of individual species.

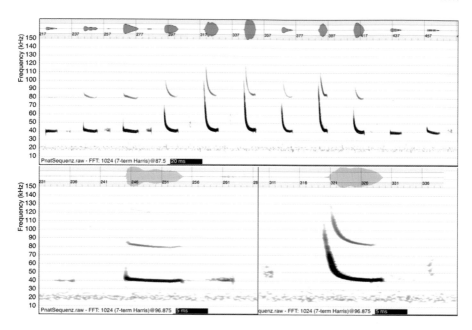

Figure 13.3   An example of a sequence of calls of Nathusius' pipistrelle. The difference between the QCF (below left) and the FM-QCF calls (below right) can be clearly seen.

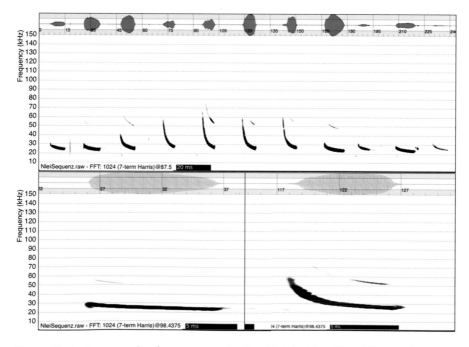

Figure 13.4   An example of a sequence of calls of Leisler's bat. The difference between the QCF (below left) and the FM-QCF calls (below right) can be clearly seen.

The sonograms in Figures 13.3 and 13.4 illustrate species that prefer foraging in open areas or edge environments. The diagrams show a representative sequence of calls. Two species are shown in detail, highlighting two types of call. The pronounced vocal plasticity of these calls is plain to see. In both cases, the bat was flying towards an edge environment, such as a hedge or wall, and changed its calls from QCF to QCF-FM. These are not only more strongly modulated but are generally of shorter duration. This is more pronounced in the nyctaloid than in the pipistrelloid species.

While the variability of QCF species is easily recognisable in single snapshots, it is a different matter in species that use FM calls. Figures 13.5 and 13.6 show sequences of a Daubenton's bat and a brown long-eared bat on a woodland edge. Here, it is not easy to recognise a change in call type. There is a change in the FM calls, but not to any great extent. While these species may use distinctly longer and less modulated calls in, for example, open areas, they generally do not change the call type as abruptly within a sequence as the nyctaloid and pipistrelloid species.

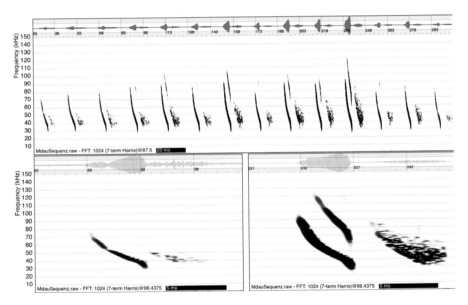

Figure 13.5   An example of a sequence of calls of Daubenton's bat. It is not possible to recognise a change in the calls through the sequence.

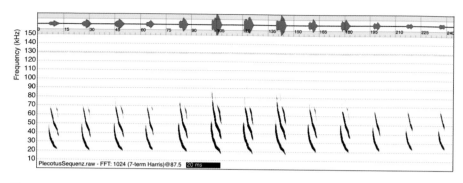

Figure 13.6 An example of a sequence of calls of brown long-eared bat. It is not possible to recognise a change in the calls through the sequence.

In sonograms, it is not uncommon to see other elements which appear to be part of the calls. This may be due to the harmonics (Figure 13.7). The second harmonic is normally at twice the frequency of the fundamental or first harmonic (the third harmonic is at three times the frequency of the first harmonic, and so on). Typically the first harmonic is the loudest, with the second, third, fourth harmonics being of decreasing intensity. However, some species depart from this general rule by actively filtering the harmonics. This is true of the long-eared (Figure 13.6) and horseshoe bats. In the case of the long-eared bats, when emitting calls through the nose, the peak intensity of the call may switch from the first to the second harmonic, resulting in two peaks of energy within the same call, and a departure from the usual rule that the peak frequency of the second harmonic is exactly twice the peak frequency of the first harmonic. Horseshoe species usually emphasise the second harmonic, and the first harmonic is much fainter and sometimes absent from sonograms. It is important to be aware that harmonics in the recordings may be spurious due to technical artefacts. For example, 'clipped' calls (those with a higher amplitude than the recording device can capture) tend to result in emphasis of the odd-numbered harmonics and loss of even-numbered harmonics (this also occurs as a matter of a course in recordings from frequency division detectors).

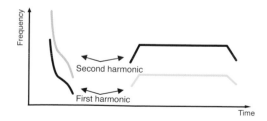

Figure 13.7 In the sonogram, further harmonics can be detected in addition to the first or fundamental harmonic.

As well as the harmonics, which may actually be part of a call, echoes might be detected (Figure 13.8). Echoes may be very clearly pronounced and come from solid objects. Diffuse echoes after the call originate from larger structures such as the foliage of trees or bushes.

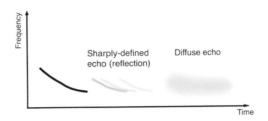

Figure 13.8   In the sonogram, it may sometimes be possible to see echoes following the bat call.

13.2 Guild structure

Each bat species will have a variety of different echolocation calls. These not only depend on the individual species, but also vary according to the task being undertaken by the bat at that moment in time. It is assumed that individual species have specialised in particular foraging techniques. These foraging techniques are primarily determined by the usable spatial structures. Species of bats will use vegetation structures to varying degrees. Those species that use vegetation in a similar way are said to belong to the same 'guild'. Bats using open environments will be in a different guild than those that are adapted to foraging in or near landscape structures. The latter guilds are strongly influenced by the horizontal heterogeneity and the vertical complexity of the habitat. These include species that forage along edge structures and that have to contend with background echoes from vegetation. Echolocation is particularly difficult for species that forage inside spaces in the vegetation, as they are subject to strong echoes from the background (Figure 13.9).

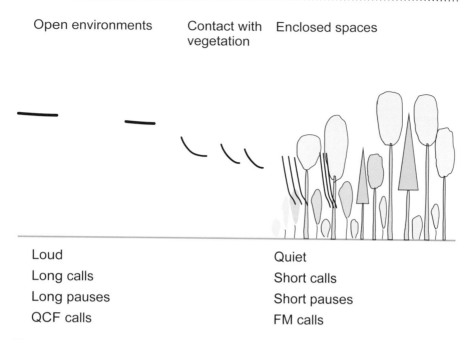

Open environments   Contact with   Enclosed spaces
                           vegetation

Loud   Quiet
Long calls   Short calls
Long pauses   Short pauses
QCF calls   FM calls

Figure 13.9   Bats are usually divided into three types of guilds, those that forage in open environments, those skirting along edge structures, and those using enclosed spaces.

When moving from an open to a closed environment, the bat, irrespective of which guild it belongs to, not only changes the type of call, but also reduces the duration of the call from over 20 ms to under 5 ms. These are adaptations to the increased overlapping of echoes from the prey with echoes from vegetation, as illustrated in Figure 13.10. The gaps between calls decrease as the background echoes increase, with the result that the call repetition rate increases. This adaptation to the environment is very flexible and generally has a significant bearing when using recordings to identify species.

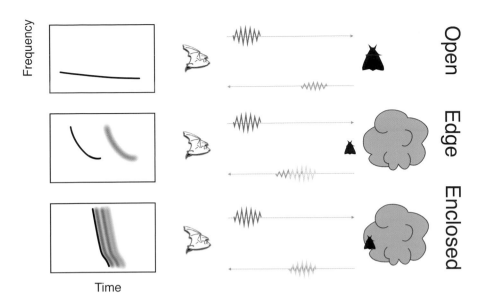

Figure 13.10 As the echolocation call echoes are increasingly overlaid with the vegetation echoes, the duration of the call decreases and the FM component increases relative to the QCF. From top to bottom, the diagram shows examples of calls for open environments, edge structures and enclosed spaces.

A good example of this adaptation is the behaviour of noctules, Leisler's and parti-coloured bats foraging in an open environment when they emit calls characterised by long, loud CF and QCF signals. The echoes from their prey will be subject to interference from echoes from vegetation and other structures. The bats respond to this by shortening their calls (Figure 13.11).

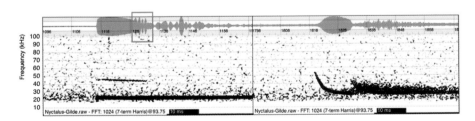

Figure 13.11 Two calls from a sequence show how a Leisler's bat solves the problem of strong echoes and overlapping with echolocation calls (left-hand diagram) by emitting shorter calls. The frame in the oscillogram in the left-hand diagram highlights the overlap with the echoes. The call in the right-hand diagram shows the call completely separated from the echoes.

Species such as the common pipistrelle and the serotine bat prefer to forage along edge structures and use FM-QCF calls. The calls are therefore usually shorter and mostly have an FM component. Nevertheless, these species will shorten their calls as the interference from echoes increases (Figure 13.12).

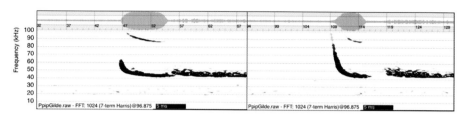

Figure 13.12   Two calls from a sequence show how the common pipistrelle deals with the problem of strong echoes and overlapping with echolocation calls (left-hand diagram) by emitting shorter calls (right-hand diagram). The call in the right-hand diagram shows the call completely separated from the echoes.

Myotis and *Plecotus* bats, which particularly favour feeding in enclosed spaces that tend to have echoes, use short FM calls to get around this problem. The calls are often very faint and so cannot be reliably detected. Moreover, some species will simply listen for the noises made by their insect prey, and so find them without echolocation, so-called passive listening. This does not mean, however, that the bat stops echolocating, as they still need this for navigation.

13.3 Feeding buzzes

Bats use particular call sequences when they are locating and catching insect prey. While the capture of prey by the bat cannot be directly proved acoustically, so-called feeding buzzes do usually indicate foraging activity, and occur in all bat species. This is more difficult to detect in *Plecotus* and *Myotis* species, however, because of their quiet calls as well as their occasional use of passive listening.

A feeding buzz is characterised by the transition from normal calls to shorter modulated calls with a more rapid repetition rate. It is usually possible to distinguish between three phases: the approach, the pursuit and the catch (Figure 13.13).

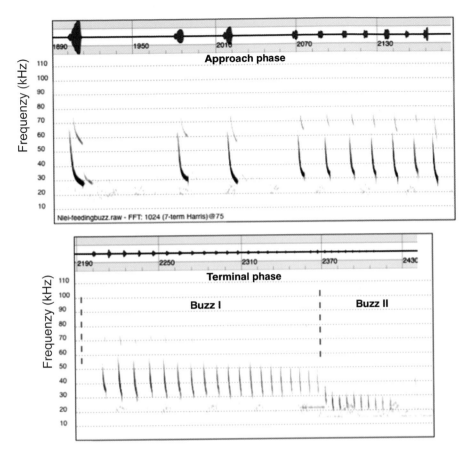

Figure 13.13   A recording of Leisler's bat shows the different phases of the feeding buzz

Feeding buzzes are an indication of hunting activity by bats (Figure 13.14). The absence of feeding buzzes does not necessarily indicate, however, that foraging is not occurring (see also Section 10.6.1, *The range of call types*).

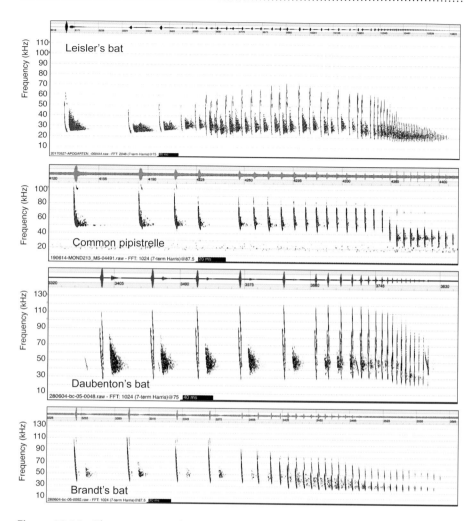

Figure 13.14   The sonograms show examples of the feeding buzzes of various species.

In a survey carried out over two years in woodland habitats (Runkel 2008), 0–31% of the sequences of the common pipistrelle were feeding buzzes, with the figure usually lying between 5% and 15%. Soprano pipistrelles had a much lower percentage of feeding buzzes. The figures for the barbastelle were 7% in many places and very occasionally between 15% and 20%. Overall, the proportion of feeding buzzes in *Myotis* species was lower. No feeding buzzes at all were recorded for the greater mouse-eared bat.

13.4 Social calls

Bats possess a very wide spectrum of social calls which serve for communication. They are particularly frequent in and around roost sites and during the mating season. With some species, such as the common pipistrelle, these social calls occur throughout the year.

The purpose of many social calls is still not clear. Although some types can be clearly attributed to mating activity, these very same calls are sometimes used during the rest of the year. Most of the other social calls are normally only rarely recorded. The purpose of social calls away from the roost is unclear, and when these do occur, it is often difficult to identify the species.

Social calls are mostly distinctive for their greater length and complexity compared to echolocation calls. This means, for example, lower frequencies, several components to the call and much more elaborate modulation (Figure 13.15). Moreover, it is also possible to observe the transition from an echolocation to a social call (Figures 13.16 and 13.17), with the result that the latter may no longer be so easily recognisable as such. A further difficulty is that the social calls of some species may strongly resemble the echolocation calls of other species. This again can make it difficult to clearly identify the species.

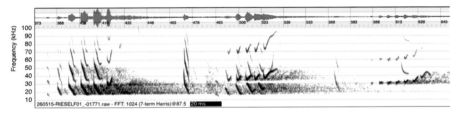

Figure 13.15   The sonogram shows a complex courtship song of Nathusius' pipistrelle, consisting of several elements.

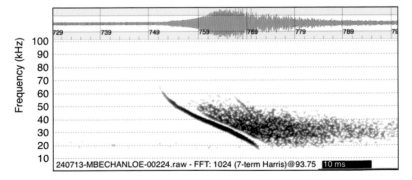

Figure 13.16   The sonogram shows a social call of Bechstein's bat which is not dissimilar to echolocation calls, albeit of a distinctly longer duration.

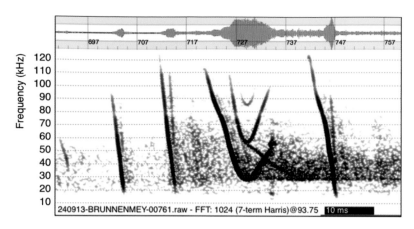

Figure 13.17 The sonogram show a social call from the courtship song of Natterer's bat. This is reminiscent of echolocation calls, but is modulated in a much more sophisticated way.

14 The physics of sound and how it is recorded

If acoustic methods are being used for surveying bats, users need to acquire a certain understanding of the properties, the propagation and the processing of sound. Without this, fundamental concepts and relationships cannot really be fully appreciated. Furthermore, when setting up devices, it is necessary to understand something of the physics of sound, such as the effects of echoes. This chapter also deals with some fundamental technical aspects of the recording and analysis of sound. Those new to bat acoustics should therefore regard this chapter as essential reading, and even experienced users should gain some new insights.

14.1 Sound

Sound is quite simply a mechanical oscillation or wave in an elastic medium. Sound needs an elastic medium, and it is this which will determine the properties of its propagation. For bats, it is a longitudinal wave travelling through the medium of air. Pressure waves oscillate in the direction of propagation, with pockets of lower pressure alternating with pockets of higher pressure. These pressure fluctuations are superimposed on the standard pressure of the air. Sound is defined by a combination of several different characteristics, which are discussed in the following sections.

14.1.1 Frequency

The frequency of sound, commonly known as pitch, describes the number of oscillations per unit of time and is measured in hertz (Hz). A sound of 20,000 Hz (20 kHz) has 20,000 oscillations per second. Sound frequencies are broken down into categories: infrasound, beyond the range of the human ear (<16 Hz), audible sound (16 Hz to 20 kHz) and inaudible ultrasound (>20 kHz). The frequency f is related directly via the speed of sound (c) to the wavelength λ:

Thus the wavelength decreases as the frequency increases. The wavelength plays an important part in the reflection and the diffraction of sound. The shorter the wavelength, the smaller will be the objects which will produce echoes. Conversely, longer wavelengths are diffracted better by bigger objects. This has implications for bats, since the detection of prey items and objects depend on the call frequency and the size of the object. Large prey items reflect sound well when the bats are calling at a low frequency. Conversely, small prey items whose size is below the wavelength hardly, if at all, reflect the ultrasound wave. As the call frequency increases, so the resolution improves. To summarise, the size of insects located very much depends on the wavelength of the bat's call.

14.1.2 Sound pressure and sound pressure level

Sound pressure in air causes fluctuations in air density. These fluctuations come about when a sound wave is propagated and interferes with the normal air pressure. Put simply, the sound pressure corresponds to the fluctuating amplitude of a sound wave. Sound pressure is measured in Pascals (Pa). As a rule, however, the sound pressure as a level in dB SPL (sound pressure level) is used in relation to $p_0 = 20\mu\ Pa = 2*10^{-5} Pa$:

$$L_p = 10 * log_{10}(\frac{\tilde{p}^2}{p_0^2})\ dB = 20 * log_{10}(\frac{\tilde{p}}{p_0})\ dB$$

The logarithmic connection between the sound pressure level and the reference pressure simplifies the representation of an otherwise very large range of values. Because of this, when considering acoustic results, relationships may have to be expressed logarithmically.

The sound pressure represents a technical measure and must not be confused with the psychoacoustic perception of volume. It is generally true, however, that an increase in the sound pressure level will also be experienced as an increase in sound volume.

14.1.3 Propagation of sound

The propagation of sound automatically leads to a reduction in sound pressure. This reduction is subject to the *1/r* rule, which means that the sound pressure will decrease in inverse proportion to the distance *r*. A doubling of the distance will reduce the sound pressure by 6 dB. This corresponds to a halving of the sound pressure level. This so-called *geometric reduction* is independent of frequency and environmental parameters. Theoretically, deviations from the *halving rule* may arise, for example, from reflections or resonance, although in practice this has not been observed to occur.

An additional factor is atmospheric dissipation, which depends on both the frequency and, importantly, the environment. The interaction of atmosphere, temperature and atmospheric humidity has a significant impact on this dissipation. Higher frequencies are more subject to dissipation than lower frequencies. Values may be from 0.5 dB/m at 20 kHz up to more than 1 dB/m at higher frequencies (Figure 14.1). This effect is obvious in foggy conditions, when low-frequency sounds are easily heard but those of a high frequency are seemingly absorbed, causing everything to sound muffled.

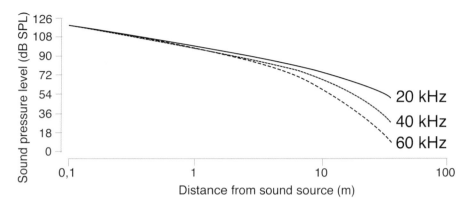

Figure 14.1   An example of sound dissipation during propagation. The sound pressure level is plotted against distance for three different frequencies. As well as the obvious geometric reduction, it can be seen that the effect of dissipation becomes more marked as the distance increases.

14.2 Microphones

The microphone is the sensor which picks up the sound. Sound pressure changes are detected, causing vibrations that are converted into electrical energy and then processed further to produce an analogue output. Filter devices, for example, can separate out undesirable components or amplify particular segments of the input. This analogue signal will then serve as the basis for digitisation.

The optimal microphone for recording bats should fulfil a number of requirements:

- It should be equally sensitive over a frequency range of 10–150 kHz. This is known as a linear frequency response.
- The signal-to-noise ratio (SNR) should be as high as possible so that quiet or weak signals can still be picked up.
- The angle of sound incidence should not affect the recording, to achieve omnidirectional reception of the sound.
- Finally, it should be resistant to all weather conditions, especially rain.

Unfortunately, this ideal microphone does not exist, so compromises on some of these requirements will have to be made.

14.2.1 Microphone types and characteristics

Microphones can be constructed according to a number of different principles. Bat detectors usually incorporate microphones with electret condensers or, more rarely, film condensers. The former are a good compromise as they are relatively sensitive to sound and are usually compact. They are, furthermore, more resistant to bad weather conditions and are omnidirectional. The great disadvantage of this type of microphone is the poor frequency response, with higher frequencies not picked up as effectively.

Film condenser microphones are very linear in their frequency response (Figure 14.2), although those with a small diameter (e.g. ⅛ inch) are not as sound-sensitive as large microphones (e.g. ½ inch). This type of microphone does tend to be very directional, and usually also very sensitive to moisture.

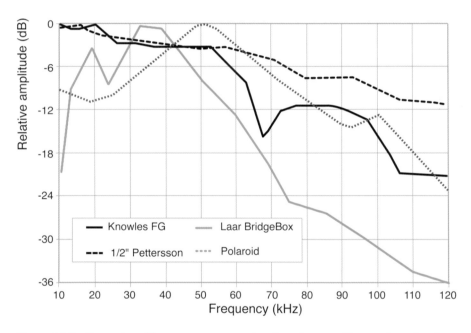

Figure 14.2 Examples of frequency response in microphones with sound coming from in front of the microphone.

Microphones which are identical in construction will not necessarily have identical properties or performance. In particular, the sound sensitivity can vary significantly (±6 dB). There may also be differences in the sound response. The film condenser microphone is especially prone to such variability, as both the tension and the evenness of the film can affect its properties. The tolerance of the components used in electret microphones will produce variations between individual microphones of the same model.

Most devices currently use an electret capsule from the Knowles FG range, which was used first in the Pettersson D240x and is also used in the batcorder. A detailed analysis of possible methods of protecting microphones from the weather can be found in Section 9.6, *Weatherproofing*.

14.2.2 Directional characteristics

The ideal acoustic sensor is equally sensitive at any frequency, irrespective of the direction of the sound. In practice, however, the structure of the microphone will always physically screen off and obscure sound from certain directions. Because of diffraction, this effect will not be identical for the whole frequency spectrum, and there will be a frequency-dependent change in the microphone's directional characteristics. Users therefore need to be aware of the directional characteristics of the microphone determined by its physical structure.

Every microphone has a characteristic polar pattern, which is the three-dimensional space surrounding the capsule where it is most sensitive to sound. The typical small electret capsules have a spherical or broad kidney-shaped polar pattern. As their size increases, this pattern changes to a kidney or even a club shape. This means that sound incidences of 30° or more will cause significant reductions of 10 or more decibels. As the frequency increases, this effect becomes greater (Figure 14.3).

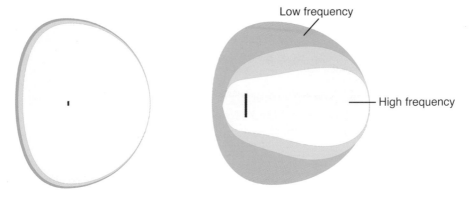

Figure 14.3 With the small electret microphone (e.g. FG) the polar pattern (the three-dimensional space surrounding the capsule where it is most sensitive to sound) tends to be spherical and is largely independent of the frequency (left). Large film microphones, in contrast, have a distinctly greater sensitivity to frequency and directionality (right).

14.2.3 Sensitivity, responsiveness and ageing

The sensitivity of a microphone is defined by the minimum sound pressure which is detected in isolation from its own background noise. The individual components of a microphone, namely resistors and condensers, are supplied with current from a power supply. They are subject to fluctuations in this current, and produce a noise.

The receiver of the sound pressure waves will only be triggered to produce a recording when a certain pressure threshold is reached. This is tantamount to a definition of the sensitivity of the microphone. With traditional microphones, this threshold is usually between 24 dB SPL (very sensitive film condenser microphones) and 30 dB SPL (electret capsule).

The properties of a microphone are not necessarily fixed, as its components are subject to an ageing process. This usually affects the sensitivity, in other words the lower threshold for sound recording. Over time, microphones can become, so to speak, hard of hearing. This effect may extend over the whole frequency range or just over individual frequency bands. In the case of film condenser microphones, it may be due to fatigue of the membrane, which may stretch or deform with use and become less receptive to high frequencies.

These ageing effects are not predictable. Some microphones seemingly never age, while others do so in a very short time. Weather (rain, frost or abrupt temperature fluctuations) and physical or mechanical factors are responsible for the ageing effects. The resulting changes in the properties of the microphone may mean that the data are no longer representative. Regular testing of the microphones is therefore essential to ensure accurate recording of bats.

The microphone testing process can be used to calibrate the detector. However, measurements of sound pressure level at ultrasonic frequencies are not as easy as they might appear, and require considerable effort. The measuring environment must be monitored, and reflections and scattering avoided. It is also important to know the precise loudspeaker frequency response. Ideally, these measurements should be supported by ultrasound-measuring microphones. The problem here is that a measuring station of this sort usually costs £10,000 or more.

14.3 Heterodyne detectors

One of the most popular methods for making bats audible to the human ear is to use a heterodyne detector, often simply known as a bat detector. Devices which work according to this principle are very affordable in comparison to other systems, costing sometimes as little as £40. Good heterodyne detectors that are suitable for professionals, however, will cost £200 or more.

This type of detector works by mixing signals. A signal with unknown frequencies has a known frequency mixed with it, which can be chosen by the user. By mixing the unknown frequency of the microphone signal with the selected frequency, sum and difference frequencies are generated. The difference frequencies are within a range audible to the human ear and are broadcast over the loudspeaker or through the headphones of the bat detector (Figure 14.4).

An example will serve to illustrate the principle. If a mixer frequency of 40 kHz is mixed with the call of a common pipistrelle of 45 kHz, the mix will generate a total of 85 kHz and an audible difference of 5 kHz. If the mixer frequency is 30 kHz, the total generated will be 75 kHz, with an audible difference, this time, of 15 kHz.

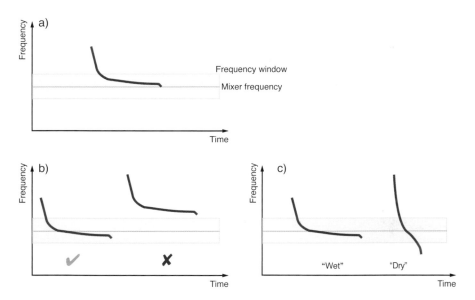

Figure 14.4   In the heterodyne process, the microphone signal is mixed with a chosen frequency. The resulting difference is audible to the human ear. In diagram (a) the signal undergoes an additional filtering ('window': heterodyne process). Species can be more readily identified by means of a frequency window (b). However, an unsuitable choice of mixer frequency may lead to some bats being missed (b). Calls with QCF/CF component sound like a wet drip, while FM calls are a dry clicking sound (c)

In order to make identification easier, and also to improve the sound quality, there may be additional filtering (super heterodyne detector). After completion of the mixing, therefore, only a part of the total frequency range is outputted. Only a narrow frequency window is used, usually with a width of ±5 kHz to ±10 kHz.

If the bat call is outside the window because of the frequency chosen, no audible signal will be heard. An unsuitable choice of frequency window can therefore lead to bats not being recorded (Figure 14.4b).

14.3.1 Identification by the human ear

The frequency adjustment of the heterodyne detector is used to find the loudest or main frequency of a call. It is possible to ascertain this by continually changing the frequency until it sounds right. The smaller the difference between the main and the mixer frequency, the deeper in pitch the sound is. With pure CF calls nothing can be heard if the frequency setting is exactly the same as the call, as the difference is zero. In general, nearly all calls have a modulated component, so that this situation can only really occur with horseshoe bats. The accuracy of the oscillator used may also vary with fluctuations in temperature, and this may lead to incorrect frequencies being shown. This error will not occur with more expensive devices.

Moreover, it is possible to hear whether a call contains constant frequency elements or is frequency modulated. The former sound 'wet', like water dripping. The reason for this is that it gives the impression of a sinusoidal sound because of the constant frequency. In contrast, frequency-modulated calls sound 'dry', or like a clicking noise (Figure 14.4c).

14.3.2 The panorama mixer

Another type of heterodyne detector is the so-called panorama mixer, which differs in having no frequency dial. The microphone signal is mixed simultaneously with several frequencies so that the whole spectrum is monitored, as in the frequency division detector discussed in Section 14.4. However, the frequency cannot be manually selected to aid identification. There are only a few models on the market which are based on this technology, and it has not established itself like the traditional heterodyne detector.

14.4 Frequency division detectors

In English-speaking countries, frequency division detectors are well established (Figure 14.5). Some users prefer them to heterodyne, since no bats will be missed with these broadband devices. The microphone signal is simply divided by ten. Every tenth oscillation is outputted, reducing a call of 40 kHz to an audible 4 kHz. The disadvantage of the process is that the amplitude of the signal is lost, although most frequency division detectors have a system for restoring this.

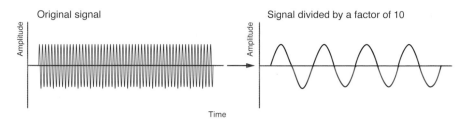

Figure 14.5 A simple explanation of the frequency division process for converting ultrasound to a range audible to the human ear.

14.4.1 Identification by human ear

The user of a frequency division detector will use a number of features to help identify the species audibly, namely the rhythm, the repetition rate and the pitch of the call. However, discriminating between frequencies is not always that easy, and a well-trained musical ear is a great advantage when working with these systems.

14.4.2 Identification by computer analysis

The converted signals can be recorded onto a normal audio recording device and then analysed on a computer. However, the call structure and various parameters tend to be less clear than with recordings from time expansion and direct sampling detectors, making species identification more challenging.

A variation on frequency division is the zero-crossing system which also carries out a zero-crossing analysis and digitisation of the divided signal. Unlike a conventional frequency division system, zero crossing simply generates data files rather than audio files which means it is not possible to listen to the recordings. This has the advantage of very small file sizes which take up little storage space and are quick to process. For many years the main zero-crossing detector has been the Anabat, though nowadays this system is included as an additional recording option on several models of direct sampling detector.

14.5 Time expansion detectors

Time expansion detectors are different in that, unlike heterodyne and time division devices, it is not possible to hear the bat calls directly. It is only after recording that the sounds are made audible at a slower rate (Figure 14.6). Usually they are slowed down by a factor of 10, so a call of 10 ms duration is slowed down to play for 100 ms. Time expansion detectors are often coupled with heterodyne devices. The calls for a set period of time are continually being stored in the background in real time. When a bat is heard on the heterodyne detector, the recording can be interrupted, and the device played in time expansion mode.

As with frequency division, the output from a time expansion detector can be recorded onto a normal recording device (ideally in WAVE format) for later analysis. The quality of the recordings is very high and the call structure is captured in higher detail than with a frequency division detector. The recordings can then be analysed at home on the computer.

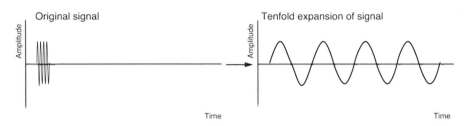

Figure 14.6   Simple diagram of the expansion process for making bat calls audible to the human ear.

The time expansion detectors work with a digital storage for buffering from 0.1 to around 3 seconds real time. When the expanded playback is outputted, the digitally stored signal is reproduced in analogue form on the headphones or line-out. There will thus be multiple analogue-to-digital/digital-to-analogue conversions of the signal, which can have a prejudicial effect on the quality. At each step an incorrect recording level could occur and cause the loss of quiet signal elements or the development of clipping effects.

14.5.1 Identification by human ear

Usually the recordings made on the time expansion detector are not used for direct identification in the field. A recording needs to be made and then listened to in slowed-down format. That wastes time in the field which could be better used for other activities. Moreover, the impression of rhythm is lost in the time expansion process. It is, however, still quite possible to recognise dry and wet calls in the time expansion mode.

14.5.2 Identification by computer analysis

Time expansion systems were the first to offer high-quality sound data for computer analysis. While you can listen to the recorded calls, the sound is mostly stored digitally and then analysed using a computer. Analysis possibilities here are the same as for direct sampled sound files.

14.6 Digitisation

Sound waves are converted to analogue signals by means of a microphone, but in order that sound can be stored and analysed on the computer, the data must be converted into discrete digital values. A reading is taken from the analogue signal at fixed intervals and the amplitude of the waves is reproduced as a range of digital values (Figure 14.7). The converter determines how many value levels are calculated (bit depth) and how often a new value is represented (the sampling rate). Audio CDs and normal wave files are written with 16 bit and 44.1 kHz.

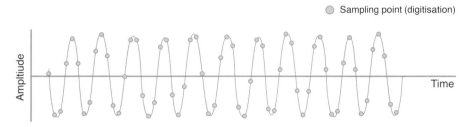

                  ● Sampling point (digitisation)

Figure 14.7   Conversion of an analogue signal to a digital signal by frequent sampling.

14.6.1 Bit depth

The bit depth indicates how many distinct values can be stored, and thus also how many values the number range contains. With a bit depth of 1, precisely $2^1 = 2$ different values can be stored, whereas an 8-bit converter can store $2^8 = 256$ values. Since a sound wave produces positive and negative amplitude levels, the value range must be halved. Therefore an 8-bit converter can reproduce 127 positive values, the zero, and 128 negative values. Modern recording systems with 16 bits will have 65,000 possible values for amplitude levels of the microphone. Time expansion detectors, in contrast, often use simpler and cheaper 8-bit converters.

With higher resolution, lower-volume signals can also be recorded. A better signal-to-noise ratio is possible. Conversely, lower resolution will tend to produce harmonics in the recordings. Higher resolution will, of course, mean greater storage requirements. However, 16 or 24 bit usually proves to be sufficient. There is no clear advantage to be gained from 32 bit, as very many of the lower bits will be taken up by system noise.

14.6.2 Sampling rates

The sampling rate indicates how often the signal is sampled for digitisation. As the sampling rate increases, so the audio-frequency which can be recorded also increases. It is vital that a sound wave is described using at least two points when it is being digitised. It follows that the maximum frequency which can theoretically be reproduced is equivalent to half the sample rate (Nyquist–Shannon sampling theorem).

Direct sampling detectors must therefore operate with a minimum sampling rate of 300 kHz (sampling the signal at least 300,000 times per second), in order to capture bat calls of up to 150 kHz. Since only two samples per wave will be captured when recording sounds at the higher end of this frequency range, the quality of the reproduction of the sound waves will tend to be unsatisfactory. If the values are by chance not at the lowest and highest points of the wave, but, for example, on the baseline, the wave will not be reproduced (Figure 14.8).

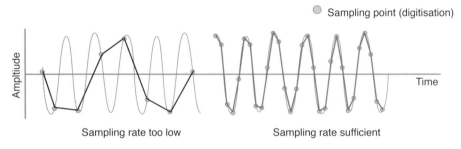

Figure 14.8   With too low a sampling rate (left) the signal is not correctly reproduced. The recorded frequencies no longer coincide with the original signal.

Table 14.1 gives an overview of typical sampling rates suitable for different bat species, and shows maximum frequencies as well as, for the purpose of analysis, good frequency ranges for various commonly used sampling rates. If the sampling rate is too low, only low frequencies are reliably recorded. It is only after a further doubling of the sampling rate that four values are available for the mapping of each sound wave. Meaningful measurements are only obtained for frequencies of between a quarter and half the sampling rate. In the case of the 300 kHz sampling rate, good reproduction will only be achieved at around 100 kHz. That is insufficient for many bat species of the genus *Myotis*. Similarly, horseshoe bats with calls above 100 kHz will not be reliably recorded. With a sampling rate of 500 kHz, in contrast, good reproduction is achieved, even at over 125 kHz. A higher resolution will mean, however, that more storage capacity will be required.

Table 14.1 Summary of typical sampling rates and resulting maximum frequencies that can be captured. A good reproduction means that there are four sampling points per wave instead of two points with the Nyquist frequency.

Sampling rate	Maximum frequency	Good reproduction	Suitability for bat calls
44.1 kHz	22.05 kHz	15 kHz	No
96 kHz	48 kHz	30 kHz	No (possibly noctule)
192 kHz	96 kHz	60 kHz	Noctules, pipistrelles
250 kHz	125 kHz	80 kHz	Noctules, pipistrelles
300 kHz	150 kHz	100 kHz	Most species (*Myotis* possibly, not all horseshoes)
500 kHz	250 kHz	150 kHz	All species

14.6.3 Storage of digital audio data

There are various formats available for the digital storage of audio data. With many of these formats, the storage format (sampling rate, bit depth) is described in the file by means of a header which precedes the actual sound data. The audio data can be stored as it is digitised or compressed. Moreover, the header may contain further information such as geodata or comments.

Typical uncompressed formats are WAVE, AIFF or RAW (batcorder). With these formats, the values are stored in direct sequence as a data stream, usually in native file format with 16 bit. Since the sampling rate is known, the data stream can be transferred again in the time-correct format to be outputted. The sound data are not modified during storage, the only exception being changes in the bit depth which are necessary for some formats.

The best-known compressed format is MP3, which was developed for the optimisation of the storage of music and language. Other compressed formats include ATRAC (MiniDisc), FLAC and WMA. While MP3 is now the most frequently used format, it does lose detail in compression. Put simply, frequency components which cannot be heard by the listener through masking are removed from the signal. This means that the compressed formats are not always suitable for call analysis, since important sound components are lost. As storage capacity is not expensive these days, it is best to avoid these formats.

When considering the usage of storage capacity at typical digitisation rates (Table 14.2), it is tempting to sample at 300 kHz instead of 500 kHz in order to make a 40% space saving. However, the result is a significantly worse frequency resolution, making automatic and manual identification of some species much harder.

Table 14.2   Storage usage at typical digitisation rates.

	16 bit, 44.1 kHz	16 bit, 300 kHz	16 bit, 500 kHz
Per second	0.089 MB	0.6 MB	1 MB
Per minute	5.34 MB	36 MB	60 MB

14.7 Representation of sound

In order to analyse calls on the computer, sound data are represented in one or more graphical formats. An oscillogram represents the data in wave form and illustrates the changes in amplitude of the recording over time. Identification of species may, however, be better carried out using the spectrogram or sonogram.

14.7.1 Representation of wave form

This representation is based on the discrete sampled values that are generated by the analogue-to-digital conversion (see also Section 14.6, *Digitisation*). The oscillogram is suitable for reviewing the recording and will usually allow the identification of calls (Figure 14.9). Furthermore, the rhythm and, if needed, the number of bats can be determined. Time measurements (call duration, inter-pulse interval) are more accurately obtained from the oscillogram than the sonogram. Beyond this, no further information regarding the identification of bats can be gleaned from the oscillogram.

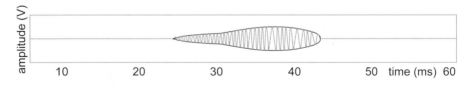

Figure 14.9   Oscillogram, plotting frequency change over time, showing a typical representation of the wave form of a recording of a bat call.

14.7.2 FFT and spectrum

An evaluation of the frequencies contained in sound is known as spectral analysis. A spectrum is usually generated by means of a Fast Fourier transform (FFT). This analysis determines the frequencies contained in the transient (continuous) signals and their energy levels. The FFT is affected by settings such as size and type of window and overlapping. The size of the window is the number of sampling values from which the FFT is calculated.

The type of window indicates how the values of the analysed segment of sound are analysed (Figure 14.10).

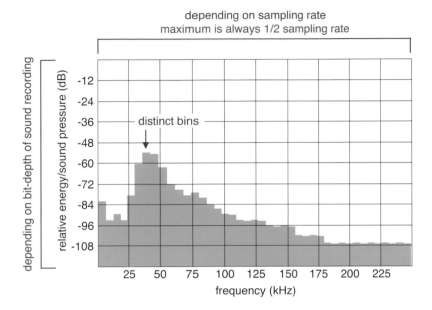

Figure 14.10   Simplified example of a spectrum. On the x-axis the frequency is displayed, with the relative energy or amplitude on the y-axis. The FFT calculates a histogram of energy distribution for frequency bins.

The size of the window has a direct effect on the time and frequency resolution of the FFT. This provides only discrete values or classes, whose breadth is determined by the size of the window. As the size of the window becomes larger, so the frequency resolution increases, and the time resolution decreases. In a recording with a sampling rate of 500 kHz and an FFT window with 1024 sampling values, the frequency resolution will be 488 Hz, and the window will cover a time segment of 2.05 ms, which is a time resolution within the range of short bat calls (Figures 14.11 and 14.12).

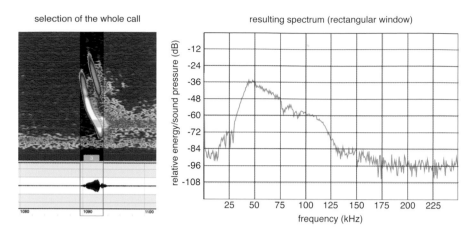

Figure 14.11   A call in waveform and as sonogram, as well as the selection of the full call required to give a proper spectral calculation. The spectrum on the right was created using an FFT at least the same size as the selection, with an additional 0 inserted to the left and right of the sound data for spectrum calculation.

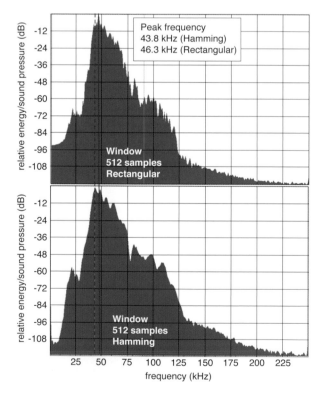

Figure 14.12   Two sets of spectral analyses. The upper one was calculated using a rectangular window with FFT size of 512 samples, the lower one using a Hamming window with otherwise the same parameters. The different weighting of the windows shows a difference in the measured peak frequency of nearly 3 kHz.

The choice of a window function will cause the margins at the beginning and end of the sound range to be degraded. There are a number of different window functions such as Hanning, Hamming or Blackman, each of which suppresses the margins in a different way (Figure 14.13).

For the calculation of a spectrum, a size of rectangular window should be selected that will include the whole of the bat call. Averaging over several windows should be avoided. Ideally, the program will pad the space before and after the sound segment with zeroes.

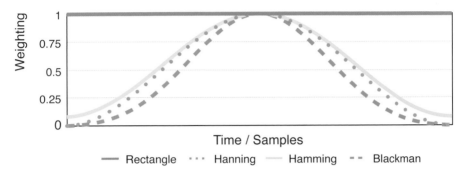

Figure 14.13   Examples of different window types for the analysis of sound signals using an FFT calculation.

These settings will ensure that there is a high-resolution sound spectrum and that the segments of the call are all evaluated equally. The use of other window functions may sometimes lead to marked differences in the sound spectrum, particularly with modulated calls, where the spectrum may show other maxima.

14.7.3 FFT and sonograms

The spectrum as a two-dimensional representation does not show the time aspects of a bat call. In order to carry out identification by type and shape of the call, the change over time of the recorded sounds must be represented. A sonogram is normally used for this purpose, with time as the x-coordinate and frequency as the y-coordinate. The volume of the frequency, or amplitude, is defined in the form of shades of grey and other colours. A sonogram is in effect a 2D projection of a 3D representation.

The basis of the representation is again FFT, and so the same framework conditions apply as previously discussed for the spectrum. Here also, the spectral structure is calculated by means of a window. The window, however, is shifted in small steps through the sound signal in order to investigate the call shape over time. In practice, this is done with an overlap, with the window being shifted only by part of its width, usually defined as a percentage of the window size. An overlap of 0% would mean a shift without overlap, whereas a 50% overlap corresponds to half the window size. A window with 1024 values and a sampling rate of 500 kHz without overlapping will result in a time resolution of 2.05 ms. With short calls, this is not sufficient to represent the development of a call shape over time. With a half-window size this gives a resolution of 1.025 ms. The window is shifted further each time by 512 values. In practice, overlaps of greater than 90% are often used (Figure 14.14). In this way, time resolutions of 0.2–0.03 ms are achieved (at a 500 kHz sampling rate). At low sampling rates, the resolution of the FFT is automatically reduced.

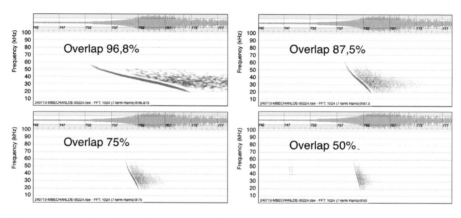

Figure 14.14   The effect of window overlapping on the time resolution. Each sonogram shows the same call, but with different time resolutions.

The window type will also determine how effectively modulated and constant-frequency call components are represented. Some windows improve the time resolution (modulated calls) whereas others increase the frequency resolution (CF calls). Only a few window types, such as the Flat top window or the seven-term Harris window, produce high resolution for both time and frequency (Figure 14.15).

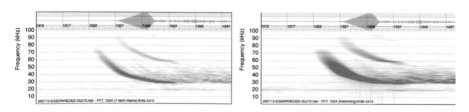

Figure 14.15   Effect of the window type on the resolution of calls. The sonogram on the left uses a seven-term Harris window and the one on the right a Hamming window.

With insufficient time or frequency resolution, identification becomes harder or even impossible. It is best to get used to employing the same settings every time. A longer overlap is preferable for short calls (<10 ms length) and a shorter overlap for longer calls.

The user must be constantly aware of the effects of the choice of window type, window size and overlap on the visual representation of the signals. A good representation will result from the selection of a big window (high frequency resolution) and a big overlap (high time resolution).

Bibliography

Abbott, I. M., Harrison, S., and Butler, F. (2012) Clutter-adaptation of bat species predicts their use of under-motorway passageways of contrasting sizes – a natural experiment. *Journal of Zoology* 287 (2): 124–132. https://doi.org/10.1111/j.1469-7998.2011.00894.x

Adams, A. M., Jantzen, M. K., Hamilton, R. M., and Fenton, M. B. (2012) Do you hear what I hear? Implications of detector selection for acoustic monitoring of bats. *Methods in Ecology and Evolution* 3 (6). https://doi.org/10.1111/j.2041-210X.2012.00244.x

Arnett, E. B., Hein, C. D., Schirmacher, M. R., Schirmacher, M. R., Huso, M. M. P., and Szewczak, J. M. (2013) Evaluating the Effectiveness of an Ultrasonic Acoustic Deterrent for Reducing Bat Fatalities at Wind Turbines. *PLoS ONE* 8 (6): e65794–11. https://doi.org/10.1371/journal.pone.0065794

Barclay, R. M. R. (1999) Bats are not birds – a cautionary note on using echolocation calls to identify bats: a comment. *Journal of Mammalogy* 80 (1): 290–296. https://doi.org/10.2307/1383229

Behr, O., Brinkmann, R., Korner-Nievergelt, F., Nagy, M., Niermann, I., Reich, M., and Simon, R. (2016) Ergebnisbericht des Forschungsvorhabens 'Reduktion des Kollisionsrisikos von Fledermäusen an Onshore-Windenergieanlagen' (RENEBAT II), pp. 1–374. https://tethys.pnnl.gov/sites/default/files/publications/Behr-et-al-2016.pdf

Behr, O., Hochradel, K., Mages, J., Korner-Nievergelt, F., Reinhard, H., Simon, R., Stiller, F., Weber, N., and Nagy, M. (2018) Bestimmung des Kollisionsrisikos von Fledermäusen an Onshore-Windenergieanlagen in der Planungspraxis (RENEBAT III), pp. 1–415. http://windbat.techfak.fau.de/Abschlussbericht/renebat-iii.pdf

Berthinussen, A. and Altringham, J. D. (2012) Do bat gantries and underpasses help bats cross roads safely? *PLoS ONE* 7 (6): e38775. https://doi.org/10.1371/journal.pone.0038775

Biscardi, S., Orprecio, J., Fenton, M. B., and Tsoar, A. (2004) Data, sample sizes and statistic affect the recognition of species of bats by their echolocation calls. *Acta Chiropterologica* 6 (2): 347–363. https://doi.org/10.3161/001.006.0212

Bosch, R., and Obrist, M. K. (2013) *BatScope – Implementation of a BioAcoustic Taxon Identification Tool.*

Brinkmann, R., Behr, O., Niermann, I., and Reich, M. (2011) *Entwicklung von Methoden zur Untersuchung und Reduktion des Kollisionsrisikos von Fledermäusen an Onshore-Windenergieanlagen.* https://www.researchgate.net/publication/285767454_Fledermausfreundliche_Betriebsalgorithmen_fur_Windenergieanlagen

Bruckner, A. (2015) Recording at water bodies increases the efficiency of a survey of temperate bats with stationary, automated detectors. *Mammalia* 80 (6): 196–199. https://doi.org/10.1515/mammalia-2014-0067

BUND, LNV, and NABU (2017) *Zur Qualität von Windenergie-Gutachten.* https://baden-wuerttemberg.nabu.de/imperia/md/content/badenwuerttemberg/positionspapiere/2017-09-07_lpk_windenergie_gutachtencheck_-_pr__sentation_langfassung.pdf

Fritsch, G. and Bruckner, A. (2014) Operator bias in software-aided bat call identification. *Ecology and Evolution* 4 (13): 2703–2713. https://doi.org/10.1002/ece3.1122

Gebhard, F., Kötteritzsch, A., Lüttmann, J., Kiefer, A., Hendler, R., and Veith, M. (2016) Fördern Arbeitshilfen die Qualität von Fachgutachten? *Naturschutz und Landschaftsplanung* 48 (6): 177ff.

Griffin, D. R. (1995) The Magic Well of bat echolocation. *Le Rhinolophe* 11: 11–15.

Griffin, D. R., Webster, F. A., and Michael, C. R. (1960) The echolocation of flying insects by bats. *Animal Behaviour* 8 (3–4): 141–154. https://doi.org/10.1016/0003-3472(60)90022-1

Hayes, J. P. (1997) Temporal variation in activity of bats and the design of echolocation-monitoring studies. *Journal of Mammalogy* 78 (2): 514–524. https://doi.org/10.2307/1382902

Hayes, J. P. (2000) Assumptions and practical considerations in the design and interpretation of echolocation-monitoring studies. *Acta Chiropterologica* 2 (2): 225–236. https://www.researchgate.net/publication/281335545_Assumptions_and_practical_considerations_in_the_design_and_interpretation_of_echolocation-monitoring_studies

Hurst, J., Balzer, S., Biedermann, M., Dietz, C., Dietz, M., Höhne, E., Karst, I., Petermann, R., Schorcht, W., Steck, C., and Brinkmann, R. (2015) Erfassungsstandards für Fledermäuse bei Windkraftprojekten in Wäldern. *Natur und Landschaft* 90 (4): 157–169. https://www.researchgate.net/deref/http%3A%2F%2Fdx.doi.org%2F10.17433%2F4.2015.50153328.157-169

Jakobsen, L., Olsen, M. N., and Surlykke, A. (2015) Dynamics of the echolocation beam during prey pursuit in aerial hawking bats. *Proceedings of the National Academy of Sciences of the United States of America* 112 (26): 8118–8123. http://doi.org/10.1073/pnas.1419943112

Jones, G. (1995) Variation in bat echolocation: implications for resource partitioning and communication. *Le Rhinolophe* 11: 53–59.

Jones, G., Vaughan, N., and Parsons, S. Acoustic identification of bats from directly sampled and time expanded recordings of vocalizations. *Acta Chiropterologica* 2 (2): 155–170.

Meschede, A., and Heller, K.-G. (2000) *Ökologie und Schutz von Fledermäusen in Wäldern.* 1. Bonn-Bad Godesberg: Bundesamt für Naturschutz.

Murray, K. L., Britzke, E. R., Hadley, B. M., and Robbins, L. W. (1999) Surveying bat communities: a comparison between mist nets and the Anabat II bat detector system. *Acta Chiropterologica* 1 (1): 105–112.

Newson, S., Ross-Smith, V., Evans, I., Harold, R., and Miller, R. (2014) Bat-monitoring: a novel approach. *British Wildlife* 25 (4): 264–269.

Obrist, M. K., Boesch, R., and Flückiger, P. F. (2004) Variability in echolocation call design of 26 Swiss bat species: consequences, limits and options for automated field identification with a synergetic pattern recognition approach. *Mammalia* 68 (4): 307–321. https://doi.org/10.1515/mamm.2004.030

O'Donnell, C. F. J. and Sedgeley, J. A. (1994) An automatic monitoring system for recording bat activity. Department of Conservation Technical Series No. 5, Department of Conservation, Wellington, New Zealand.

O'Farrell, M. J. and Gannon, W. L. (1999) A comparison of acoustic versus capture techniques for the inventory of bats. *Journal of Mammology* 80 (1): 24–30. https://doi.org/10.2307/1383204

Parsons, S., Boonman, A. M. and Obrist, M. K. (2000) Advantages and disadvantages of techniques for transforming and analyzing chiropteran echolocation calls. *Journal of Mammology* 81 (4): 92–98. https://doi.org/10.1644/1545-1542(2000)081<0927:AADOTF>2.0.CO;2

Pye, D. (1993) Is fidelity futile? The true signal is illusory, especially with ultrasound. *Bioacoustics* 4: 271–286. https://doi.org/10.1080/09524622.1993.10510438

Runkel, V. (2008) *Mikrohabitatnutzung syntoper Waldfledermaeuse*, Dissertation.

Russo, D. and Voigt, C. C. (2016) The use of automated identification of bat echolocation calls in acoustic monitoring: A cautionary note for a sound analysis. *Ecological Indicators* 66: 598–602. https://doi.org/10.1016/j.ecolind.2016.02.036

Rydell, J., Nyman, S., Eklöf, J., Jones, G., and Russo, D. (2017) Testing the performances of automated identification of bat echolocation calls: A request for prudence. *Ecological Indicators* 78: 416–420. https://doi.org/10.1016/j.ecolind.2017.03.023

Schnitzler, H.-U. and Kalko, E. K. V. (2001) Echolocation by insect-eating bats. *BioSience* 51 (7): 557–569. https://doi.org/10.1641/0006-3568(2001)051[0557:EBIEB]2.0.CO;2

Waters, D. A. and Walsh, A. L. (1994) The influence of bat detector brand on the quantitative estimation of bat activity. *Bioacoustics* 5 (3): 205–221. https://doi.org/10.1080/09524622.1994.9753245

Wilkinson, G. S., and Bradbury, J. W. (1988) Radiotelemetry: Techniques and Analyses. In: Kunz, T. H. (Hrsg.): *Radiotelemetry: Techniques and Analyses*. Smithsonian Institution, pp. 105–124.

Zingg, P. E. (1990) Akustische Artidentifikation von Fledermäusen (Mammalia: Chiroptera) in der Schweiz. *Revue suisse Zool.* 97 (2): 263–294. https://doi.org/10.5962/bhl.part.92388

Index

Locators to plans and tables are in *italics*.